SpringerBriefs in Applied Sciences and Technology

SpringerBriefs present concise summaries of cutting-edge research and practical applications across a wide spectrum of fields. Featuring compact volumes of 50 to 125 pages, the series covers a range of content from professional to academic.

Typical publications can be:

- A timely report of state-of-the art methods
- An introduction to or a manual for the application of mathematical or computer techniques
- A bridge between new research results, as published in journal articles
- A snapshot of a hot or emerging topic
- An in-depth case study
- A presentation of core concepts that students must understand in order to make independent contributions

SpringerBriefs are characterized by fast, global electronic dissemination, standard publishing contracts, standardized manuscript preparation and formatting guidelines, and expedited production schedules.

On the one hand, **SpringerBriefs in Applied Sciences and Technology** are devoted to the publication of fundamentals and applications within the different classical engineering disciplines as well as in interdisciplinary fields that recently emerged between these areas. On the other hand, as the boundary separating fundamental research and applied technology is more and more dissolving, this series is particularly open to trans-disciplinary topics between fundamental science and engineering.

Indexed by EI-Compendex, SCOPUS and Springerlink.

More information about this series at https://link.springer.com/bookseries/8884

Luigi Romano

Advanced Brush Tyre Modelling

Springer

Luigi Romano
Mechanics and Maritime Science
Chalmers University of Technology
Gothenburg, Västra Götalands Län, Sweden

ISSN 2191-530X ISSN 2191-5318 (electronic)
SpringerBriefs in Applied Sciences and Technology
ISBN 978-3-030-98434-2 ISBN 978-3-030-98435-9 (eBook)
https://doi.org/10.1007/978-3-030-98435-9

Series Standard cover

This Springer imprint is published by the registered company Springer Nature Switzerland AG
The registered company address is: Gewerbestrasse 11, 6330 Cham, Switzerland

Preface

Due to their unique properties and their almost exclusive role, tyres represent a fundamental topic in Vehicle Dynamics. It is perhaps superfluous to say that an enormous amount of work has been spent on the subject, which has attracted many researchers from both the academic and industrial worlds. Traditionally addressed from the perspective of classic contact mechanics, the interaction between the tyre and the road has progressively aroused the interest of a broader community, including control engineers and vehicle dynamicists. Amongst them, the focus is probably on developing reliable models which may be used in simulations, controller design and friction and state estimation. In this context, it becomes even more crucial to understand the main phenomena which determine the transient response of the tyre, regarded as a nonlinear system of forces and moments. Many of these phenomena may be studied within the theoretical framework provided by the brush theory.

Therefore, the idea behind this monograph is to explain the nonstationary dynamics of the tyre from the perspective of the brush models. The ultimate goal is to provide a high-level description relating the tyre characteristics (the forces and moments acting upon the tyre) and the so-called slip inputs. According to the brush models, this may be achieved using some fundamental intuitions from the in-depth analysis of the local phenomena occurring inside the contact patch.

The results presented in this book should hopefully inspire researchers and engineers in the development of algorithms and functions based on an intuitive understanding of the tyre-road contact mechanics. Students pursuing a Master's or a PhD in control, with research interests in Vehicle Dynamics, may also approach the topic at an advanced level.

The monograph is organised as follows: Chap. 1 provides a high-level description of the tyre as a nonlinear system. The slip inputs are defined and the fundamental assumptions on the functions used to describe the steady-state tyre characteristics are outlined. Chapter 2 introduces the governing equations of the brush tyre models and states the main assumptions on which the theory is grounded. In Chap. 3, the classic theory of steady-state generation of tyre forces and moment is reviewed. The analysis is propaedeutic to address the classic and exact nonstationary theories presented in Chaps. 4 and 5, respectively. Finally, Chap. 6 is dedicated to some pragmatic transient models which may be effectively used in simulations: the single contact point and the two-regime formulation.

Gothenburg, Sweden Luigi Romano

Acknowledgements

In writing this monograph, there are a few people to whom I would like to express my gratitude. First of all, my supervisor Fredrik Bruzelius for encouraging me and patiently supporting my work on tyre modelling. Also, Bengt Jacobson as a second supervisor for the valuable guidance provided during these first two years of my PhD.

Many thanks also go to Francesco Timpone, Massimo Guiggiani and Alex O'Neill for inspiring my research with several passionate discussions. Tyres are a very intriguing world indeed.

Finally, I would like to thank my family and my beloved Yulia for the immense love they spoil me every day.

Contents

Nomenclature

Forces and Moments

$\boldsymbol{F}_t$	Tangential force vector (N)
$\boldsymbol{F}_{t0}$	Initial conditions for the tangential force vector (N)
F_x, F_y	Longitudinal and lateral tyre forces (N)
F_{x0}, F_{y0}	Initial conditions for the longitudinal and lateral tyre forces (N)
F_z	Vertical force (N)
M_z	Self-aligning moment (Nm)
$\boldsymbol{q}_t$	Shear stress vector (Nm^{-2})
q_t	Total tangential shear stress (Nm^{-2})
q_x, q_y	Longitudinal and lateral shear stress (Nm^{-2})
$\boldsymbol{q}_t^{(a)}$	Shear stress vector in the adhesion region (Nm^{-2})
$q_x^{(a)}$, $q_y^{(a)}$	Longitudinal and lateral shear stress in the adhesion region (Nm^{-2})
$\boldsymbol{q}_t^{(s)}$	Shear stress vector in the sliding region (Nm^{-2})
$q_x^{(s)}$, $q_y^{(s)}$	Longitudinal and lateral shear stress in the sliding region (Nm^{-2})
$\boldsymbol{q}_t^{-}$	Stationary shear stress vector (Nm^{-2})
q_x^{-}, q_y^{-}	Stationary longitudinal and lateral shear stress (Nm^{-2})
$\boldsymbol{q}_t^{-}$	Transient shear stress vector (Nm^{-2})
q_x^{-}, q_y^{-}	Transient longitudinal and lateral shear stress (Nm^{-2})
q_z	Vertical pressure (Nm^{-2})
q_z^*	Vertical pressure peak (Nm^{-2})
$q_z^{(\eta)}$	Restriction of the vertical pressure (Nm^{-2})

Displacements

$\boldsymbol{u_t}$	Tangential deflection vector of the bristle (m)
u_x, u_y	Longitudinal and lateral deflection of the bristle (m)
$\boldsymbol{u}_{\boldsymbol{t}}^{(\mathrm{a})}$	Tangential deflection vector of the bristle in the adhesion region (m)
$u_x^{(\mathrm{a})}$, $u_y^{(\mathrm{a})}$	Longitudinal and lateral deflection in the adhesion region (m)
$\boldsymbol{u}_{\boldsymbol{t}}^{(\mathrm{s})}$	Tangential deflection vector of the bristle in the sliding region (m)
$u_x^{(\mathrm{s})}$, $u_y^{(\mathrm{s})}$	Longitudinal and lateral deflection in the sliding region (m)
$\boldsymbol{u}_{\boldsymbol{t}}^{-}$	Stationary tangential deflection vector of the bristle (m)
u_x^{-}, u_y^{-}	Stationary longitudinal and lateral deflection (m)
$\boldsymbol{u}_{\boldsymbol{t}}^{+}$	Transient tangential deflection vector of the bristle (m)
u_x^{+}, u_y^{+}	Transient longitudinal and lateral deflection (m)
$\boldsymbol{u}_{\boldsymbol{t}0}$	Initial tangential deflection vector of the bristle (IC) (m)
u_{x0}, u_{y0}	Initial longitudinal and lateral deflection (IC) (m)
s	Travelled distance (m)
$\boldsymbol{x}$	Coordinate vector (m)
x, y, z	Longitudinal and lateral coordinates (m)
$\boldsymbol{x}_0$	Initial data vector (ID) (m)
x_0, y_0	Initial longitudinal and lateral data (ID) (m)
$\boldsymbol{\xi}$	Local coordinate vector (m)
ξ, η, ζ	Alternative longitudinal, lateral and vertical coordinates (m)
$\boldsymbol{\delta_t}$	Tyre carcass tangential displacement vector (m)
δ_x, δ_y	Tyre carcass longitudinal and lateral displacements (m)

Speeds

$\boldsymbol{V}_{\mathrm{r}}$	Rolling velocity (ms^{-1})
V_{r}	Rolling speed (ms^{-1})
$\boldsymbol{V}$	Velocity of the wheel hub (ms^{-1})
V_x, V_y	Longitudinal and lateral speed of the wheel hub (ms^{-1})
$\boldsymbol{V}_C$	Velocity of the actual contact point (ms^{-1})
V_{Cx}, V_{Cy}	Longitudinal and lateral speed of the actual contact point (ms^{-1})
$\boldsymbol{V}_{\mathrm{s}}$	Sliding velocity (ms^{-1})
$\boldsymbol{V}_{\mathrm{s}}'$	Transient sliding velocity (ms^{-1})
$V_{\mathrm{s}x}$, $V_{\mathrm{s}y}$	Longitudinal and lateral sliding speed (ms^{-1})
$V_{\mathrm{s}x}'$, $V_{\mathrm{s}y}'$	Transient longitudinal and lateral sliding speed (ms^{-1})
Ω	Angular speed of the tyre ($\mathrm{rad\,s}^{-1}$)
$\bar{\boldsymbol{v}}_{\boldsymbol{t}}$	Nondimensional tangential velocity field (–)
$\bar{v}_x$, $\bar{v}_y$	Longitudinal and lateral components of the nondimensional velocity field (–)
$\bar{\boldsymbol{v}}_{\mathrm{s}}$	Nondimensional micro-sliding velocity (–)

$\bar{v}_{sx}$, $\bar{v}_{sy}$	Longitudinal and lateral nondimensional micro-sliding speeds (–)
$\bar{\boldsymbol{v}}_{\mathscr{S}}^{(\hat{\nu})}$	Particular nondimensional velocity of the sliding edge (–)
$\bar{\boldsymbol{v}}_{\Sigma}^{(\hat{\nu})}$	Particular nondimensional velocity of the travelling edge (–)
$\bar{\boldsymbol{v}}_{\partial\mathscr{P}}$	Nondimensional velocity of the boundary of the contact patch (–)
$\dot{\psi}$	Steering speed (rad s^{-1})
$\boldsymbol{\omega}$	Angular velocity of the tyre (rad s^{-1})
ω_x, ω_y, ω_z	Angular speeds of the tyre around the x, y and z axes (rad s^{-1})

Slip Parameters

χ_γ, χ_ψ	Camber and turn ratio (–)
ε_γ	Camber reduction factor (–)
α	Slip angle (geometrical lateral slip) (–)
α'	Transient slip angle (transient geometrical lateral slip) (–)
γ	Camber angle (geometrical spin) (–)
$\boldsymbol{\kappa}$	Practical translational slip vector (–)
κ_x, κ_y	Practical longitudinal and lateral slip (–)
$\boldsymbol{\kappa}'$	Transient practical translational slip vector (–)
κ_x', κ_y'	Transient practical longitudinal and lateral slip (–)
$\boldsymbol{\sigma}$	Theoretical translational slip vector (–)
σ	Theoretical total translational slip (–)
σ_x, σ_y	Theoretical longitudinal and lateral slip (–)
$\boldsymbol{\sigma}'$	Transient theoretical slip vector (–)
σ_x', σ_y'	Transient theoretical longitudinal and lateral slip (–)
σ^*	Local critical slip (–)
σ^{cr}	Global critical slip (–)
$\tilde{\sigma}^{cr}$	Equivalent critical slip (–)
σ_χ	Transient critical slip (–)
φ	Rotational slip or spin parameter (m^{-1})
φ^{cr}	Critical spin (m^{-1})
φ_γ, φ_ψ	Camber and turn spin parameters (m^{-1})

Rotation Matrices and Tensors

$\mathbf{A}_\varphi$, $\mathbf{A}_{\varphi_\gamma}$, $\mathbf{A}_{\varphi_\psi}$	Spin, camber spin and turn spin tensors (m^{-1})
$\mathbf{R}_{\varphi_\gamma}$	Camber spin rotation matrix (–)
$\mathbf{R}_{\varphi_\psi}$	Turning spin rotation matrix (–)
$\boldsymbol{\Phi}_{\varphi_\gamma}$	Transition matrix for camber spin (–)
$\boldsymbol{\Phi}_{\varphi_\psi}$	Transition matrix for turn spin (–)

Geometric Parameters

a, b	Contact patch semilength and semiwidth (m)
c_{r}	Lateral offset due to camber (m)
t_{p}	Pneumatic trail (m)
R_{r}	Rolling radius (m)
R_{δ}	Deformed radius of the tyre (m)
R_{γ}, R_{ψ}	Cambering and turning radius (m)
$\boldsymbol{\Lambda}_{\sigma}$	Matrix of theoretical slip relaxation lengths (m)
$\tilde{\boldsymbol{\Lambda}}_{\sigma}$	Matrix of generalised theoretical slip relaxation lengths (m)
$\boldsymbol{\Lambda}'_{\sigma}$	Matrix of enhanced theoretical slip relaxation lengths (m)
$\boldsymbol{\Lambda}'_{\varphi}$	Matrix of enhanced theoretical spin relaxation lengths (m)
$\boldsymbol{\Lambda}'_{M\sigma}$	Matrix of enhanced aligning theoretical slip relaxation lengths (m)
$\boldsymbol{\Lambda}_{\kappa}$	Matrix of practical slip relaxation lengths (m)
$\tilde{\boldsymbol{\Lambda}}_{\kappa}$	Matrix of generalised practical slip relaxation lengths (m)
$\boldsymbol{\Lambda}'_{\kappa}$	Matrix of enhanced practical slip relaxation lengths (m)
$\boldsymbol{\Lambda}'_{M\kappa}$	Matrix of enhanced aligning practical slip relaxation lengths (m)
$\lambda_{x\alpha}, \lambda_{y\alpha}$	Geometrical lateral relaxation lengths (m)
$\tilde{\lambda}_{x\alpha}, \tilde{\lambda}_{y\alpha}$	Generalised geometrical lateral relaxation lengths (m)
$\lambda'_{x\alpha}, \lambda'_{y\alpha}$	Enhanced geometrical lateral relaxation lengths (m)
$\lambda'_{x\gamma}, \lambda'_{y\gamma}$	Enhanced geometrical spin (camber) relaxation lengths (m)
$\lambda'_{M\alpha}$	Enhanced aligning lateral relaxation lengths (m)
$\lambda'_{M\gamma}$	Enhanced aligning geometrical spin (camber) relaxation lengths (m)
λ_{ad}	Adhesion length (m)
χ_{λ}	Relaxation ratio (–)

Stiffnesses and Compliances

$\mathbf{K}_{t}$	Matrix of the bristle tangential stiffnesses (Nm^{-3})
k	Bristle stiffness (isotropic tyre) (Nm^{-3})
$\mathbf{C}_{\sigma}$	Matrix of theoretical slip stiffnesses (N)
$\tilde{\mathbf{C}}_{\sigma}$	Matrix of generalised theoretical slip stiffnesses (N)
$\mathbf{C}'_{\sigma}$	Matrix of enhanced theoretical slip stiffnesses (Nm^{-1})
$\mathbf{C}_{\kappa}$	Matrix of practical slip stiffnesses (N)
$\tilde{\mathbf{C}}_{\kappa}$	Matrix of generalised practical slip stiffnesses (N)
$\mathbf{C}'_{\kappa}$	Matrix of enhanced practical slip stiffnesses (Nm^{-1})
C_{σ}	Slip stiffness (isotropic tyre) (N)
C_{φ}	Spin stiffness (isotropic tyre) (Nm)
$C_{x\alpha}, C_{y\alpha}$	Geometrical lateral slip stiffnesses (N)
$C'_{x\alpha}, C'_{y\alpha}$	Enhanced geometrical lateral slip stiffnesses (Nm^{-1})
$C_{x\gamma}, C_{y\gamma}$	Longitudinal and lateral geometrical (camber) spin stiffnesses (N)

$C'_{x\gamma}$, $C'_{y\gamma}$ Enhanced longitudinal and lateral geometrical (camber) spin stiffnesses (Nm^{-1})
$C_{M\alpha}$ Aligning geometrical lateral slip stiffnesses (Nm)
$C'_{M\alpha}$ Enhanced aligning geometrical lateral slip stiffnesses (N)
$C_{M\gamma}$ Aligning geometrical spin (camber) stiffnesses (Nm)
$C'_{M\gamma}$ Enhanced aligning geometrical spin (camber) stiffnesses (N)
$C_{M\varphi}$ Aligning spin stiffnesses (Nm^2)
$C'_{M\varphi}$ Enhanced aligning spin stiffnesses (Nm)
$\mathbf{C}_{\varphi}$ Matrix of theoretical spin stiffnesses (Nm)
$\tilde{\mathbf{C}}_{\varphi}$ Matrix of generalised theoretical spin stiffnesses (Nm)
$\mathbf{C}_{\varphi}$ Matrix of theoretical spin stiffnesses (reduced) (Nm)
$\mathbf{C}'_{\varphi}$ Enhanced matrix of theoretical spin stiffnesses (reduced) (N)
$\mathbf{C}_{M\sigma}$ Matrix of theoretical aligning slip stiffnesses (Nm)
$\tilde{\mathbf{C}}_{M\sigma}$ Matrix of generalised theoretical aligning slip stiffnesses (Nm)
$\mathbf{C}'_{M\sigma}$ Matrix of enhanced theoretical aligning slip stiffnesses (N)
$\mathbf{C}_{M\kappa}$ Matrix of practical aligning slip stiffnesses (Nm)
$\tilde{\mathbf{C}}_{M\kappa}$ Matrix of generalised practical aligning slip stiffnesses (Nm)
$\mathbf{C}'_{M\kappa}$ Matrix of enhanced practical aligning slip stiffnesses (N)
$\mathbf{C}_{M\varphi}$ Matrix of theoretical aligning spin stiffnesses (Nm^2)
$\tilde{\mathbf{C}}_{M\varphi}$ Matrix of generalised theoretical aligning spin stiffnesses (Nm^2)
$\mathbf{C}'$ Matrix of tyre carcass stiffnesses (Nm^{-1})
$\mathbf{S}'$ Matrix of tyre carcass compliances (Nm^{-1})

Friction Parameters

μ Friction coefficient (–)
$\tilde{\mu}$ Equivalent friction coefficient (–)

Functions and Operators

∇_t Tangential gradient (m^{-1})
$x_{\mathscr{A}}$, $\xi_{\mathscr{A}}$ Explicit parametrisations of an adhesion edge (–)
$x_{\mathscr{L}}$, $y_{\mathscr{L}}$, $\xi_{\mathscr{L}}$ Explicit parametrisations of a leading edge (–)
$x_{\mathscr{N}}$, $y_{\mathscr{N}}$, $\xi_{\mathscr{N}}$ Explicit parametrisations of a neutral edge (–)
$x_{\mathscr{S}}$, $\xi_{\mathscr{S}}$ Explicit parametrisations of a sliding edge (–)
x_{Σ}, ξ_{Σ} Explicit parametrisations of a travelling edge (–)
$x_{\mathscr{T}}$, $y_{\mathscr{T}}$, $\xi_{\mathscr{T}}$ Explicit parametrisations of a trailing edge (–)
$\Gamma(\cdot)$ Gamma function (m^2)
$\Sigma(\cdot)$ Sigma function (m)
$\boldsymbol{\Psi}(\cdot)$ Vector-valued psi function (m)

$\Psi_x(\cdot)$, $\Psi_y(\cdot)$	Longitudinal and lateral psi functions (m)
$\gamma_{\mathscr{S}}$	Implicit parametrisation of a sliding edge (–)
γ_{Σ}	Implicit parametrisation of a travelling edge (–)
$\tan\check{\alpha}(\cdot,\cdot)$, $\sin\check{\gamma}(\cdot,\cdot)$	Geometrical sliding functions (–)
$\tan\hat{\alpha}(\cdot,\cdot)$, $\sin\hat{\gamma}(\cdot,\cdot)$	Geometrical slip functions (–)
$\check{\boldsymbol{\kappa}}(\cdot,\cdot)$	Practical sliding functions (–)
$\hat{\boldsymbol{\kappa}}(\cdot,\cdot)$	Practical slip functions (–)
$\check{\boldsymbol{\sigma}}(\cdot,\cdot)$, $\check{\varphi}(\cdot,\cdot)$	Theoretical sliding functions (–, m^{-1})
$\hat{\boldsymbol{\sigma}}(\cdot,\cdot)$, $\hat{\varphi}(\cdot,\cdot)$	Theoretical slip functions (–, m^{-1})

Sets

$\mathscr{P}$	Contact patch (m^2)
$\mathscr{P}_0$	Initial conditions for the contact patch (m^2)
$\mathscr{P}^{(\mathrm{a})}$, $\mathscr{P}^{(\mathrm{s})}$	Adhesion and sliding region (m^2)
$\mathscr{P}_0^{(\mathrm{a})}$, $\mathscr{P}_0^{(\mathrm{s})}$	Initial conditions for the adhesion and sliding region (m^2)
$\mathscr{P}^{(\eta)}$	Restriction of the contact patch (m^2)
$\mathring{\mathscr{P}}$	Interior of the contact patch $\mathscr{P}$ (m^2)
$\mathring{\mathscr{P}}_0$	Initial condition for the interior of the contact patch (m^2)
$\mathring{\mathscr{P}}^{(\mathrm{a})}$, $\mathring{\mathscr{P}}^{(\mathrm{s})}$	Interior of the adhesion and sliding region (m^2)
$\mathring{\mathscr{P}}_0^{(\mathrm{a})}$, $\mathring{\mathscr{P}}_0^{(\mathrm{s})}$	Initial conditions for the interior of the adhesion and sliding region (m^2)
$\partial\mathscr{P}$	Boundary of the contact patch $\mathscr{P}$ (m)
$\partial\mathscr{P}_0$	Initial condition for the boundary of the contact patch (m)
$\partial\mathscr{P}^{(\mathrm{a})}$, $\partial\mathscr{P}^{(\mathrm{s})}$	Boundary of the adhesion and sliding region (m)
$\mathscr{A}$	Adhesion edge (m)
$\mathscr{L}$	Leading edge (m)
$\mathscr{L}_0$	Initial condition for the leading edge (m)
$\mathscr{N}$	Neutral edge (m)
$\mathscr{S}$	Sliding edge (m)
$\mathscr{S}_0$	Initial condition for the sliding edge (m)
$\mathscr{T}$	Trailing edge (m)
$\mathbb{R}_{\geq 0}$	Set of positive real numbers (including 0) (–)
$\mathbb{R}_{>0}$	Set of strictly positive real numbers (excluding 0) (–)
Π	Road plane (m^2)

Chapter 1
The Tyre as a Nonlinear System

Abstract At a high level of description, the tyre may be thought of as a nonlinear dynamical system, which produces certain outputs, often referred to as *tyre characteristics*, when subjected to opportune inputs. This interpretation allows defining some fundamental quantities that contribute to determining both the steady-state and the transient response of the tyre. Amongst these, the slip variables play the most important role. In steady-state conditions, the tyre characteristics may be described as real analytic functions of the slips, which may be defined in three main different ways. The Jacobians of the steady-state characteristics with respect to the slip variables are often called matrices of generalised slip stiffnesses. The local properties of the steady-state tyre characteristics may be deduced from the entries of these matrices.

The scope of this chapter is to introduce a high-level description of the main physical entities involved in the tyre-road interaction problem. The perspective is rather general and considers both transient and steady-state phenomena. The fundamental concept introduced in the chapter is that the behaviour of the tyre may be generally captured by a dynamic nonlinearity. As illustrated in Fig. 1.1, the idea is to relate the output of the tyre as a nonlinear system, namely the tyre characteristics, to the fundamental inputs which determine its behaviour. The dashed lines in Fig. 1.1 correspond to the inputs and outputs that are not directly considered in this chapter. More specifically, the considered inputs are the slip and spin variables, whilst the outputs are the tangential forces and the self-aligning moment.

1.1 Reference Frame and Wheel Velocities

To properly define the tyre inputs, it is first necessary to introduce the kinematic parameters that describe the relative motion between the tyre-wheel system and the road. In this book, the contact between the tyre and the road is studied according to the reference frame illustrated in Fig. 1.2 [1]: the x-axis is directed as the longitudinal direction of motion, the z-axis points downward and the y-axis is oriented so to have

L. Romano, *Advanced Brush Tyre Modelling*,
SpringerBriefs in Applied Sciences and Technology,
https://doi.org/10.1007/978-3-030-98435-9_1

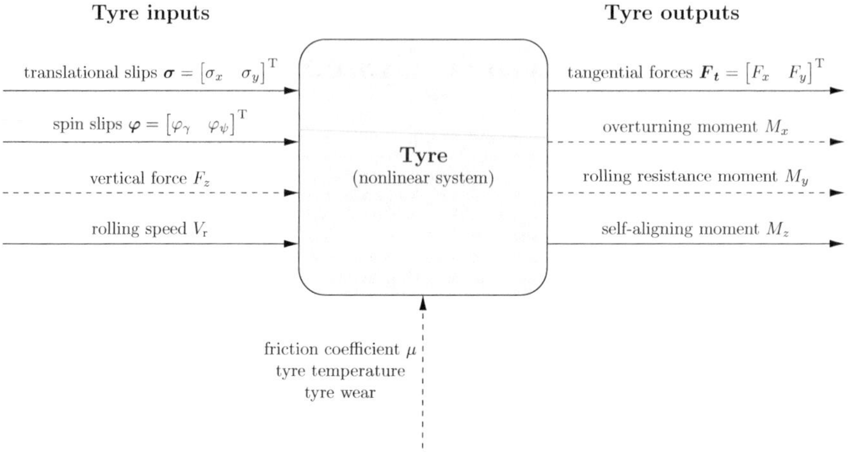

Fig. 1.1 The tyre as a nonlinear system. The dashed lines correspond to the inputs and outputs which are not directly considered in this chapter

Fig. 1.2 Tyre-wheel system and reference frame

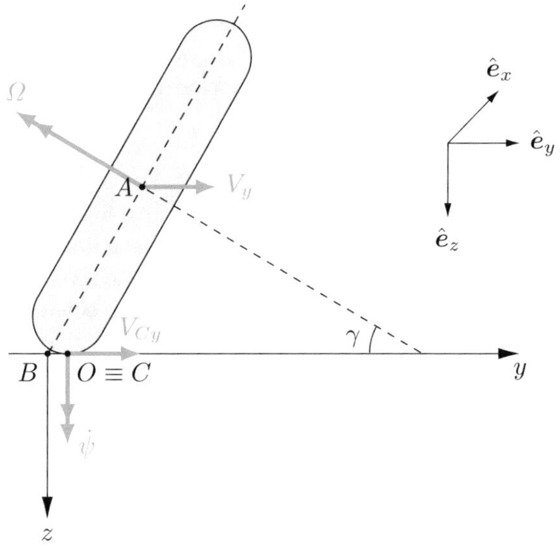

a right-handed system; the origin usually coincides with the *virtual contact point* B, defined by the intersection of the two middle planes of the tyre and the road plane $\Pi = \{\boldsymbol{x} \in \mathbb{R}^3 \mid z = 0\}$, which is assumed to be completely flat in Fig. 1.2. Other choices for the reference frame are, however, possible [2, 3].

The tyre characteristics $\boldsymbol{F}_t$, M_z and the vertical force F_z are traditionally concentrated in the virtual contact point B. The *actual contact point* or *contact centre* C, on the other hand, lies on the road plane, displaced laterally to B by a small offset $c_{\mathrm{r}}(\gamma)$ resulting from the inclination of the wheel due to a finite camber angle γ (Fig. 1.3).

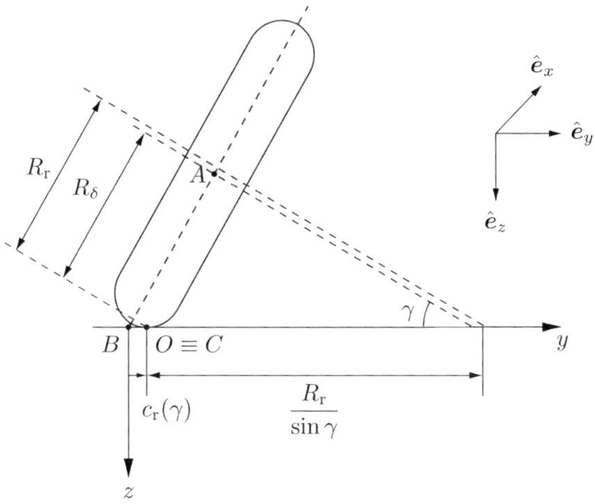

Fig. 1.3 A small offset $c_r(\gamma)$ between the virtual contact point B and the actual contact point C is caused by a finite camber angle γ

In *pure rolling* conditions, that is $F_x = F_y = M_z = 0$, the point C travels with the rolling velocity of the tyre $\boldsymbol{V}_r$ defined below [3].

Another reference point, attached to the contact patch $\mathscr{P}$ (it may coincide with its centre when $\mathscr{P}$ is symmetric), is instead denoted by O. In normal operating conditions, O may be displaced by the actual contact point C by a quantity which corresponds to the deformation of the tyre carcass (Fig. 1.4). When the tyre carcass is undeformed, or the tyre-wheel system is assumed to be rigid, $O \equiv C$.

The centre A of the wheel hub and the point C travel with total velocities given by

$$\boldsymbol{V}_A = \boldsymbol{V} = \begin{bmatrix} V_x \\ V_y \end{bmatrix}, \tag{1.1a}$$

$$\boldsymbol{V}_C = \begin{bmatrix} V_{Cx} \\ V_{Cy} \end{bmatrix}. \tag{1.1b}$$

Furthermore, the tyre-wheel system has an angular velocity $\boldsymbol{\omega}$ reading

$$\boldsymbol{\omega} = \begin{bmatrix} \omega_x \\ \omega_y \\ \omega_z \end{bmatrix} = \begin{bmatrix} \dot{\gamma} \\ -\Omega \cos\gamma \\ \dot{\psi} - \Omega \sin\gamma \end{bmatrix}, \tag{1.2}$$

where γ is the camber angle, $\Omega > 0$ is the angular speed of the wheel hub around its axis and $\dot{\psi}$ is the spin angular speed of the wheel around the vertical axis, as shown in Fig. 1.3. Clearly, when $\gamma\dot{\psi} = 0$, the velocities of the wheel hub centre and the actual contact point coincide, that is $\boldsymbol{V}_A \equiv \boldsymbol{V}_C \equiv \boldsymbol{V}$. The *macro-sliding velocity* or simply *sliding velocity*, denoted by $\boldsymbol{V}_s$, represents the difference between

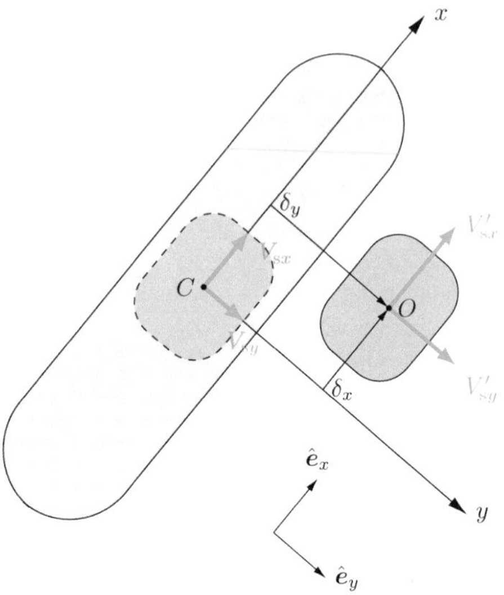

Fig. 1.4 In normal operating conditions, the point O may be displaced from C by a quantity which coincides with the deflection of the tyre carcass

the velocity of C and the rolling velocity of the tyre. It may be deduced using the kinematic relationships for a rigid body and reads

$$\boldsymbol{V}_{\mathrm{s}} = \begin{bmatrix} V_{\mathrm{s}x} \\ V_{\mathrm{s}y} \end{bmatrix} \triangleq \boldsymbol{V}_C - \boldsymbol{V}_{\mathrm{r}} = \begin{bmatrix} V_{Cx} - V_{\mathrm{r}} \\ V_{Cy} \end{bmatrix}, \tag{1.3}$$

where $\boldsymbol{V}_{\mathrm{r}} = V_{\mathrm{r}}\hat{\boldsymbol{e}}_x = \Omega R_{\mathrm{r}}\hat{\boldsymbol{e}}_x$ is the above-mentioned rolling velocity, V_{r} is the scalar *rolling speed* and $R_{\mathrm{r}} = R_{\mathrm{r}}(\gamma)$ is called *rolling radius*. As also shown in Fig. 1.3, R_{r} is usually greater than the tyre deformed radius R_δ and needs to be determined experimentally for different values of the camber angle γ. As already discussed, the actual contact point C and the contact patch centre O do not coincide due to the flexibility of the tyre carcass, which may deform of a quantity

$$\boldsymbol{\delta}_t = \begin{bmatrix} \delta_x \\ \delta_y \end{bmatrix}. \tag{1.4}$$

As a consequence, in transient conditions, the actual sliding velocity of the contact patch centre becomes

$$\boldsymbol{V}'_{\mathrm{s}} = \begin{bmatrix} V'_{\mathrm{s}x} \\ V'_{\mathrm{s}y} \end{bmatrix} \triangleq \boldsymbol{V}_{\mathrm{s}} + \dot{\boldsymbol{\delta}}_t. \tag{1.5}$$

The velocity $\boldsymbol{V}'_{\mathrm{s}}$ is called in this book *transient sliding speed*, for reasons which will be clarified in Chap. 4.

The kinematic quantities introduced so far constitute the basis for the definition of the so-called slip variables, which fully capture the steady-state response of the tyre. In transient operating conditions, on the other hand, the kinematic quantities presented in this section complement the slip inputs in determining the response of the tyre. At least, if the time is used as independent variable, as explained in Chap. 2.

1.2 Slip Variables

Probably, the most important role in the generation of tyre forces and moment is played by the so-called *slip variables*. Traditionally, different definitions have been proposed in the literature for the slip variables. In particular, it is possible to distinguish between *theoretical slips*, *practical slips* and *geometrical slips*.

1.2.1 Theoretical Slip Variables

The theoretical slips represent the *natural* slip variables in the description of the tyre-road contact mechanics. Their definition is, indeed, provided directly by the partial differential equations (PDEs) governing the tyre-road interaction, as shown in Chap. 2 .

Specifically, the translational slip $\boldsymbol{\sigma}$ and the spin slip φ may be introduced as [1, 4]

$$\boldsymbol{\sigma} = \begin{bmatrix} \sigma_x \\ \sigma_y \end{bmatrix} \triangleq -\frac{\mathbf{V}_\mathrm{s}}{V_\mathrm{r}} = -\frac{1}{V_\mathrm{r}}\begin{bmatrix} V_{\mathrm{s}x} \\ V_{\mathrm{s}y} \end{bmatrix} \triangleq -\frac{1}{V_\mathrm{r}}\begin{bmatrix} V_{Cx} - V_\mathrm{r} \\ V_{Cy} \end{bmatrix}, \tag{1.6a}$$

$$\varphi \triangleq -\frac{\omega_z}{V_\mathrm{r}} - \frac{1}{R_\mathrm{r}}\varepsilon_\gamma \sin\gamma = -\frac{1}{V_\mathrm{r}}\Big[\dot{\psi} - (1-\varepsilon_\gamma)\Omega\sin\gamma\Big] = -\frac{\dot{\psi}}{V_\mathrm{r}} + \frac{1}{R_\mathrm{r}}(1-\varepsilon_\gamma)\sin\gamma, \tag{1.6b}$$

where σ_x and σ_y are called theoretical longitudinal and lateral slip, respectively. It may be understood that the translational slips are nondimensional quantities, whereas the total spin slip φ has the same dimension of a curvature. In turn, it may be decomposed in the camber and turn spins φ_γ and φ_ψ defined, respectively, as

$$\varphi_\gamma \triangleq \frac{1}{R_\mathrm{r}}(1-\varepsilon_\gamma)\sin\gamma \triangleq \frac{1}{R_\gamma}, \qquad \varphi_\psi \triangleq -\frac{\dot{\psi}}{V_\mathrm{r}} \triangleq -\frac{1}{R_\psi}. \tag{1.7}$$

From Eq. (1.7), it may be inferred that the total spin $\varphi = \varphi_\gamma + \varphi_\psi = 1/R_\gamma - 1/R_\psi$ actually represents the sum of two different signed curvatures [5]. Generally speaking, these two curvatures produce different effects on the generation of tyre forces and moment and must be treated separately [8]. However, when they are sufficiently small, their contribution can be fairly merged using the total spin φ. This approach is very common in practice.

It is also useful to introduce the camber and turn spin ratios as follows:

$$\chi_\gamma \triangleq \frac{\varphi_\gamma}{\varphi}, \qquad \chi_\psi \triangleq \frac{\varphi_\psi}{\varphi}. \tag{1.8}$$

Hereafter, the complete set of theoretical slip variables is denoted in this chapter by $(\boldsymbol{\sigma}, \boldsymbol{\varphi})$, where $\boldsymbol{\varphi} = (\varphi_\gamma, \varphi_\psi)$. If the problem can be simplified by considering the total spin in place of the two separate contributions in Eq. (1.7), then the theoretical slips are indicated using the triplet $(\boldsymbol{\sigma}, \varphi)$.

1.2.2 Practical Slip Variables

Similarly to the theoretical translational slips $\boldsymbol{\sigma}$, it is also customary to define the so-called practical translational slips $\boldsymbol{\kappa}$ as

$$\boldsymbol{\kappa} = \begin{bmatrix} \kappa_x \\ \kappa_y \end{bmatrix} \triangleq -\frac{\boldsymbol{V}_s}{V_{Cx}} \equiv \frac{\boldsymbol{\sigma}}{1-\sigma_x}, \tag{1.9}$$

where in this case κ_x and κ_y are referred to as practical longitudinal and lateral slips. The inverse relationship between the vectors $\boldsymbol{\sigma}$ and $\boldsymbol{\kappa}$ is clearly given by

$$\boldsymbol{\sigma} = \frac{\boldsymbol{\kappa}}{1+\kappa_x}. \tag{1.10}$$

The theoretical and practical slips are, therefore, equivalent for small longitudinal slips σ_x.

The practical lateral slip variable κ_y may also be expressed as a function of the slip angle α as

$$\kappa_y = \tan\alpha. \tag{1.11}$$

When the tyre-road interaction is described using the practical translational slips, the tuple $(\boldsymbol{\kappa}, \boldsymbol{\varphi})$ may be used, with $\boldsymbol{\varphi}$ again defined as in Eq. (1.7). In this context, the variables $(\boldsymbol{\kappa}, \boldsymbol{\varphi})$ are referred to as *practical slips* in general, even though $\boldsymbol{\varphi}$ still collects the theoretical spin slip. Furthermore, if the contributions of the camber and turn spins can be condensed in the total spin, the triplet $(\boldsymbol{\kappa}, \varphi)$ is used.

1.2.3 Geometrical Slip Variables

Sometimes, the geometrical slip variables are used. These collect the slip angle α and the camber angle γ:

$$\alpha = \arctan\left(-\frac{V_{sy}}{V_{Cx}}\right) \equiv \arctan \kappa_y, \tag{1.12a}$$

$$\gamma = \arcsin\left(\frac{\varphi_\gamma R_r}{1-\varepsilon_\gamma}\right). \tag{1.12b}$$

It should be noticed that the slip angle α is nondimensional, as the practical slip κ_y. On the other hand, whilst the camber angle γ is nondimensional, the camber spin has the dimension of a curvature.

When the tyre-road interaction is studied using the geometrical slips, the tuple $(\kappa_x, \alpha, \gamma, \varphi_\psi)$ may be used. These variables are generally referred to as *geometrical slips*, even though κ_x is practical and φ_ψ is theoretical. Alternatively, when the turn slip is negligible, the problem may be fully described using the triplet $(\kappa_x, \alpha, \gamma)$.

1.2.4 Transient Slip Variables

To explain the presence of nonstationary phenomena within the framework of the brush theory, the effect of a compliant tyre carcass may be considered. According to the definition of transient sliding velocity $\boldsymbol{V}'_s$, the *transient slips* are defined as

$$\boldsymbol{\sigma}' \triangleq \frac{\boldsymbol{V}'_s}{V_r} = \boldsymbol{\sigma} - \frac{\dot{\boldsymbol{\delta}}_t}{V_r}, \tag{1.13}$$

and

$$\boldsymbol{\kappa}' \triangleq \frac{\boldsymbol{V}'_s}{V_{Cx}} = \boldsymbol{\kappa} - \frac{\dot{\boldsymbol{\delta}}_t}{V_{Cx}}, \tag{1.14}$$

and

$$\tan\alpha' \triangleq \frac{V'_{sy}}{V_{Cx}} = \tan\alpha - \frac{\dot{\delta}_y}{V_{Cx}}. \tag{1.15}$$

The slips above are referred to as transient since they capture the dynamics of the tyre carcass. The transformations between the transient slip variables are formally identical to those between their conventional counterparts.

1.3 General Tyre-Slip Relationships

From a purely mathematical perspective, a very general way of describing a tyre in transient conditions is using a nonlinear system of *differential-algebraic equations* (DAEs) of the type[1]

$$\boldsymbol{f}\Big(V_{\rm r}, \boldsymbol{\sigma}, \boldsymbol{\varphi}, \boldsymbol{F}_t, \dot{\boldsymbol{F}}_t, F_z, M_z, \dot{M}_z\Big) = \boldsymbol{0}, \tag{1.16}$$

or

$$\boldsymbol{f}\Big(V_{Cx}, V_{\rm r}, \boldsymbol{\kappa}, \boldsymbol{\varphi}, \boldsymbol{F}_t, \dot{\boldsymbol{F}}_t, F_z, M_z, \dot{M}_z\Big) = \boldsymbol{0} \tag{1.17}$$

if the practical slip variables are used in place of the theoretical ones. Alternatively, another possibility is to employ the geometrical slips:

$$\boldsymbol{f}\Big(V_{Cx}, V_{\rm r}, \kappa_x, \alpha, \gamma, \varphi_\psi, \boldsymbol{F}_t, \dot{\boldsymbol{F}}_t, F_z, M_z, \dot{M}_z\Big) = \boldsymbol{0}. \tag{1.18}$$

It is worth noticing that the implicit relationships (1.16), (1.17) and (1.18) may (and actually do) contain additional arguments that influence the tyre dynamics, for example, the inflation pressure and the tyre temperature [6, 7]. However, for the present analysis, these quantities are not directly relevant.

The fundamental problem in tyre dynamics is to derive, when possible, explicit relationships between the derivatives of the tangential tyre forces and moments in Eqs. (1.16), (1.17) and (1.18) and the other motion-related quantities. Often, this is a hard task, and only simple approximated differential relationships may be found.

On the other hand, Eqs. (1.19), (1.20) and (1.21) notably simplify in steady-state conditions. In this case, the rolling and longitudinal speeds $V_{\rm r}$ and V_{Cx} have usually a minor influence upon the generation of tyre forces and moment, which motivates to write

$$\boldsymbol{g}(\boldsymbol{\sigma}, \boldsymbol{\varphi}, \boldsymbol{F}_t, F_z, M_z) = \boldsymbol{0}, \tag{1.19}$$

or

$$\boldsymbol{g}(\boldsymbol{\kappa}, \boldsymbol{\varphi}, \boldsymbol{F}_t, F_z, M_z) = \boldsymbol{0}, \tag{1.20}$$

or

$$\boldsymbol{g}\big(\kappa_x, \alpha, \gamma, \varphi_\psi, \boldsymbol{F}_t, F_z, M_z\big) = \boldsymbol{0}, \tag{1.21}$$

respectively.

In steady-state conditions, it is often possible to derive explicit nonlinear expressions describing the tyre forces and moment $(\boldsymbol{F}_t, M_z)$ as a function of the slips

---

[1] The self-aligning moment M_z may be generally referred to the points B, C or O. In Eqs. (1.19), (1.20) and (1.21), in the remaining of the chapter and also in Chaps. 3 and 6, the moment computed with respect to the contact patch origin O (or the virtual contact point when the tyre carcass is undeformed) is considered, that is $M_z = M_z^{(O)}$.

$(\boldsymbol{\sigma}, \boldsymbol{\varphi})$, $(\boldsymbol{\kappa}, \boldsymbol{\varphi})$ or $(\kappa_x, \alpha, \gamma, \varphi_\psi)$ and for constant values of the vertical load F_z. In this context, the slips are usually treated as independent variables. However, depending on the specific nature of the problem at hand, the tyre forces and moments and the slips may be both regarded as static inputs or nonlinearities in turn.

1.4 Steady-State Relationships

As previously mentioned, steady-state operating conditions of the tyre imply the existence of static mappings between the forces and the slips. The vertical force is often regarded as a parameter, and, therefore, the explicit dependency on F_z is hereafter omitted for the sake of simplicity.

1.4.1 Tyre-Slip Steady-State Relationships

When the slips are regarded as independent variables, a system of explicit relationships may be derived describing the triplet $(\boldsymbol{F}_t, M_z)$ as a function of the slip and spin variables[2]:

$$\boldsymbol{F}_t = \boldsymbol{F}_t(\boldsymbol{\sigma}, \boldsymbol{\varphi}), \tag{1.22a}$$

$$M_z = M_z(\boldsymbol{\sigma}, \boldsymbol{\varphi}), \tag{1.22b}$$

or

$$\boldsymbol{F}_t = \boldsymbol{F}_t(\boldsymbol{\kappa}, \boldsymbol{\varphi}), \tag{1.23a}$$

$$M_z = M_z(\boldsymbol{\kappa}, \boldsymbol{\varphi}), \tag{1.23b}$$

or

$$\boldsymbol{F}_t = \boldsymbol{F}_t\big(\kappa_x, \alpha, \gamma, \varphi_\psi\big), \tag{1.24a}$$

$$M_z = M_z\big(\kappa_x, \alpha, \gamma, \varphi_\psi\big). \tag{1.24b}$$

The three systems are the equivalent explicit forms of Eqs. (1.19), (1.20) and (1.21), depending on the specific choice of slip variables (theoretical, practical or geometrical). Usually, such relationships are found directly from some assumed or postulated model—either physical or empirical—and may, thus, be thought of as a natural set of nonlinear functions describing the tyre-road interaction in steady-state conditions.

---

[2] To be formally correct, different symbols should be used to denote the mappings and the related physical quantities. However, for ease of interpretation, the notation is often mixed in the book. Some exceptions may be found in Chap. 6.

In this context, Eqs. (1.22), (1.23) and (1.24) are generally studied with respect to their smoothness properties. Usually, these functions are continuous, and at least locally C^1. This is an important property, which allows defining several useful quantities.

1.4.1.1 Theoretical slip and spin stiffnesses

A first-order Taylor approximation of the mappings (1.22) in the neighbourhood of some point $(\bar{\boldsymbol{\sigma}}, \bar{\boldsymbol{\varphi}})$ yields

$$\boldsymbol{F}_t(\boldsymbol{\sigma}, \boldsymbol{\varphi}) \approx \boldsymbol{F}_t(\bar{\boldsymbol{\sigma}}, \bar{\boldsymbol{\varphi}}) + \nabla_{\boldsymbol{\sigma}} \boldsymbol{F}_t(\bar{\boldsymbol{\sigma}}, \bar{\boldsymbol{\varphi}})^{\mathrm{T}}(\boldsymbol{\sigma} - \bar{\boldsymbol{\sigma}}) + \nabla_{\boldsymbol{\varphi}} \boldsymbol{F}_t(\bar{\boldsymbol{\sigma}}, \bar{\boldsymbol{\varphi}})^{\mathrm{T}}(\boldsymbol{\varphi} - \bar{\boldsymbol{\varphi}}), \tag{1.25a}$$

$$M_z(\boldsymbol{\sigma}, \boldsymbol{\varphi}) \approx M_z(\bar{\boldsymbol{\sigma}}, \bar{\boldsymbol{\varphi}}) + \nabla_{\boldsymbol{\sigma}} M_z(\bar{\boldsymbol{\sigma}}, \bar{\boldsymbol{\varphi}})^{\mathrm{T}}(\boldsymbol{\sigma} - \bar{\boldsymbol{\sigma}}) + \nabla_{\boldsymbol{\varphi}} M_z(\bar{\boldsymbol{\sigma}}, \bar{\boldsymbol{\varphi}})^{\mathrm{T}}(\boldsymbol{\varphi} - \bar{\boldsymbol{\varphi}}). \tag{1.25b}$$

The role played by the slip stiffness matrices, that is the Jacobians appearing in Eqs. (1.25a), (1.25b), is of vital importance in vehicle dynamics, to the point that even their entries have specific names. Indeed, it is customary to define the following quantities:

$$\tilde{\mathbf{C}}_{\boldsymbol{\sigma}}(\boldsymbol{\sigma}, \boldsymbol{\varphi}) = \begin{bmatrix} \tilde{C}_{x\sigma_x}(\boldsymbol{\sigma}, \boldsymbol{\varphi}) & \tilde{C}_{x\sigma_y}(\boldsymbol{\sigma}, \boldsymbol{\varphi}) \\ \tilde{C}_{y\sigma_x}(\boldsymbol{\sigma}, \boldsymbol{\varphi}) & \tilde{C}_{y\sigma_y}(\boldsymbol{\sigma}, \boldsymbol{\varphi}) \end{bmatrix} \triangleq \begin{bmatrix} \dfrac{\partial F_x(\boldsymbol{\sigma}, \boldsymbol{\varphi})}{\partial \sigma_x} & \dfrac{\partial F_x(\boldsymbol{\sigma}, \boldsymbol{\varphi})}{\partial \sigma_y} \\ \dfrac{\partial F_y(\boldsymbol{\sigma}, \boldsymbol{\varphi})}{\partial \sigma_x} & \dfrac{\partial F_y(\boldsymbol{\sigma}, \boldsymbol{\varphi})}{\partial \sigma_y} \end{bmatrix} = \nabla_{\boldsymbol{\sigma}} \boldsymbol{F}_t(\boldsymbol{\sigma}, \boldsymbol{\varphi})^{\mathrm{T}}, \tag{1.26a}$$

$$\tilde{\mathbf{C}}_{\boldsymbol{\varphi}}(\boldsymbol{\sigma}, \boldsymbol{\varphi}) = \begin{bmatrix} \tilde{C}_{x\varphi_\gamma}(\boldsymbol{\sigma}, \boldsymbol{\varphi}) & \tilde{C}_{x\varphi_\psi}(\boldsymbol{\sigma}, \boldsymbol{\varphi}) \\ \tilde{C}_{y\varphi_\gamma}(\boldsymbol{\sigma}, \boldsymbol{\varphi}) & \tilde{C}_{y\varphi_\psi}(\boldsymbol{\sigma}, \boldsymbol{\varphi}) \end{bmatrix} \triangleq \begin{bmatrix} \dfrac{\partial F_x(\boldsymbol{\sigma}, \boldsymbol{\varphi})}{\partial \varphi_\gamma} & \dfrac{\partial F_x(\boldsymbol{\sigma}, \boldsymbol{\varphi})}{\partial \varphi_\psi} \\ \dfrac{\partial F_y(\boldsymbol{\sigma}, \boldsymbol{\varphi})}{\partial \varphi_\gamma} & \dfrac{\partial F_y(\boldsymbol{\sigma}, \boldsymbol{\varphi})}{\partial \varphi_\psi} \end{bmatrix} = \nabla_{\boldsymbol{\varphi}} \boldsymbol{F}_t(\boldsymbol{\sigma}, \boldsymbol{\varphi})^{\mathrm{T}}, \tag{1.26b}$$

$$\tilde{\mathbf{C}}_{M\boldsymbol{\sigma}}(\boldsymbol{\sigma}, \boldsymbol{\varphi}) = \begin{bmatrix} \tilde{C}_{M\sigma_x}(\boldsymbol{\sigma}, \boldsymbol{\varphi}) & \tilde{C}_{M\sigma_y}(\boldsymbol{\sigma}, \boldsymbol{\varphi}) \end{bmatrix} \triangleq - \begin{bmatrix} \dfrac{\partial M_z(\boldsymbol{\sigma}, \boldsymbol{\varphi})}{\partial \sigma_x} & \dfrac{\partial M_z(\boldsymbol{\sigma}, \boldsymbol{\varphi})}{\partial \sigma_y} \end{bmatrix} = -\nabla_{\boldsymbol{\sigma}} M_z(\boldsymbol{\sigma}, \boldsymbol{\varphi})^{\mathrm{T}}, \tag{1.26c}$$

$$\tilde{\mathbf{C}}_{M\boldsymbol{\varphi}}(\boldsymbol{\sigma}, \boldsymbol{\varphi}) = \begin{bmatrix} \tilde{C}_{M\varphi_\gamma}(\boldsymbol{\sigma}, \boldsymbol{\varphi}) & \tilde{C}_{M\varphi_\psi}(\boldsymbol{\sigma}, \boldsymbol{\varphi}) \end{bmatrix} \triangleq \begin{bmatrix} \dfrac{\partial M_z(\boldsymbol{\sigma}, \boldsymbol{\varphi})}{\partial \varphi_\gamma} & \dfrac{\partial M_z(\boldsymbol{\sigma}, \boldsymbol{\varphi})}{\partial \varphi_\psi} \end{bmatrix} = \nabla_{\boldsymbol{\varphi}} M_z(\boldsymbol{\sigma}, \boldsymbol{\varphi})^{\mathrm{T}}. \tag{1.26d}$$

More specifically, the entries in the theoretical slip stiffness matrix $\tilde{\mathbf{C}}_{\boldsymbol{\sigma}}$ (Eq. (1.26a)) have the same dimension of a force, since they multiply the dimensionless vector $\boldsymbol{\sigma}$. The terms $\tilde{C}_{x\sigma_x}$ and $C_{y\sigma_y}$ are referred to as principal longitudinal and lateral slip stiffness, respectively. The entries and $C_{y\sigma_x}$ and $C_{x\sigma_y}$ are instead called cross longitudinal and lateral slip stiffnesses. The second matrix $\tilde{\mathbf{C}}_{\varphi}$, appearing in Eq. (1.26b) and called spin stiffness matrix, collects the longitudinal and lateral camber spin stiffnesses $\tilde{C}_{x\varphi_\gamma}$, $\tilde{C}_{y\varphi_\gamma}$ and the turn spin stiffnesses $\tilde{C}_{x\varphi_\psi}$, $\tilde{C}_{y\varphi_\psi}$. In Eq. (1.26c), the stiffnesses $\tilde{C}_{M\sigma_x}$ and $\tilde{C}_{M\sigma_y}$ in the aligning slip stiffness matrix $\tilde{\mathbf{C}}_{M\sigma}$ have again the dimension of a moment and are called aligning longitudinal and lateral slip stiffness. Finally, in Eq. (1.26d), the aligning camber and turn spin stiffnesses $\tilde{C}_{M\varphi_\gamma}$ and $\tilde{C}_{M\varphi_\psi}$ in $\tilde{\mathbf{C}}_{M\varphi}$ are expressed in N m^2.

All the stiffnesses introduced so far are referred to as *theoretical*, since they represent the partial derivatives of the tyre forces and moment with respect to the theoretical slip and spin variables. They are also *generalised*, being themselves functions of the quantities $(\boldsymbol{\sigma}, \boldsymbol{\varphi})$. However, since the slips are usually small, it is a common practice to linearise the tyre forces and moment in the neighbourhood of the origin. This leads to the definition of the even more fundamental conventional stiffnesses, which may be derived directly from Eqs. (1.26a), (1.26b), (1.26c), (1.26d), as

$$\mathbf{C}_{\sigma} = \begin{bmatrix} C_{x\sigma_x} & C_{x\sigma_y} \\ C_{y\sigma_x} & C_{y\sigma_y} \end{bmatrix} \triangleq \tilde{\mathbf{C}}_{\sigma}(\mathbf{0},\mathbf{0}) = \nabla_{\sigma} \boldsymbol{F}_t(\boldsymbol{\sigma},\boldsymbol{\varphi})^{\mathrm{T}} \Big|_{\boldsymbol{\sigma}=\mathbf{0},\boldsymbol{\varphi}=\mathbf{0}}, \tag{1.27a}$$

$$\mathbf{C}_{\varphi} = \begin{bmatrix} C_{x\varphi_\gamma} & C_{x\varphi_\psi} \\ C_{y\varphi_\gamma} & C_{y\varphi_\psi} \end{bmatrix} \triangleq \tilde{\mathbf{C}}_{\varphi}(\mathbf{0},\mathbf{0}) = \nabla_{\varphi} \boldsymbol{F}_t(\boldsymbol{\sigma},\boldsymbol{\varphi})^{\mathrm{T}} \Big|_{\boldsymbol{\sigma}=\mathbf{0},\boldsymbol{\varphi}=\mathbf{0}}, \tag{1.27b}$$

$$\mathbf{C}_{M\sigma} = \begin{bmatrix} C_{M\sigma_x} & C_{M\sigma_y} \end{bmatrix} \triangleq \tilde{\mathbf{C}}_{M\sigma}(\mathbf{0},\mathbf{0}) = -\nabla_{\sigma} M_z(\boldsymbol{\sigma},\boldsymbol{\varphi})^{\mathrm{T}} \Big|_{\boldsymbol{\sigma}=\mathbf{0},\boldsymbol{\varphi}=\mathbf{0}}, \tag{1.27c}$$

$$\mathbf{C}_{M\varphi} = \begin{bmatrix} C_{M\varphi_\gamma} & C_{M\varphi_\psi} \end{bmatrix} \triangleq \tilde{\mathbf{C}}_{M\varphi}(\mathbf{0},\mathbf{0}) = \nabla_{\varphi} M_z(\boldsymbol{\sigma},\boldsymbol{\varphi})^{\mathrm{T}} \Big|_{\boldsymbol{\sigma}=\mathbf{0},\boldsymbol{\varphi}=\mathbf{0}}. \tag{1.27d}$$

Another common simplification may be introduced by noticing that, for $\boldsymbol{\varphi} \simeq \mathbf{0}$, there is no formal difference between the camber and turn spin. Therefore, the latter two variables are more conveniently replaced by the total spin $\varphi = \varphi_\gamma + \varphi_\psi$, and the matrices defined in Eqs. (1.26b) and (1.26d) are approximated as

$$\tilde{\mathbf{C}}_{\varphi}(\boldsymbol{\sigma},\boldsymbol{\varphi}) \approx \tilde{\mathbf{C}}_{\varphi}(\boldsymbol{\sigma},\varphi) = \begin{bmatrix} \tilde{C}_{x\varphi}(\boldsymbol{\sigma},\varphi) \\ \tilde{C}_{y\varphi}(\boldsymbol{\sigma},\varphi) \end{bmatrix} \triangleq \frac{\partial \boldsymbol{F}_t(\boldsymbol{\sigma},\varphi)}{\partial \varphi}, \tag{1.28a}$$

$$\tilde{\mathbf{C}}_{M\varphi}(\boldsymbol{\sigma},\boldsymbol{\varphi}) \approx \tilde{C}_{M\varphi}(\boldsymbol{\sigma},\varphi) \triangleq \frac{\partial M_z(\boldsymbol{\sigma},\varphi)}{\partial \varphi}. \tag{1.28b}$$

The corresponding conventional stiffnesses may be computed as

$$\mathbf{C}_\varphi = \begin{bmatrix} C_{x\varphi} \\ C_{y\varphi} \end{bmatrix} \triangleq \tilde{\mathbf{C}}_\varphi(\mathbf{0}, 0) = \left.\frac{\partial \boldsymbol{F}_t(\boldsymbol{\sigma}, \varphi)}{\partial \varphi}\right|_{\boldsymbol{\sigma}=\mathbf{0}, \varphi=0}, \tag{1.29a}$$

$$C_{M\varphi} \triangleq \tilde{C}_{M\varphi}(\mathbf{0}, 0) = \left.\frac{\partial M_z(\boldsymbol{\sigma}, \varphi)}{\partial \varphi}\right|_{\boldsymbol{\sigma}=\mathbf{0}, \varphi=0}. \tag{1.29b}$$

It may be understood that, in the case of small spin slips, the nonlinear relationships describing the tyre forces and moment simplify considerably, since the number of involved variables diminishes by one. Indeed, the problem is completely determined by the mappings (1.22a), (1.22b), where the independent variables reduce to the triplet $(\boldsymbol{\sigma}, \varphi)$.

1.4.1.2 Practical Slip Stiffnesses

When the practical slips $(\boldsymbol{\kappa}, \varphi)$ are used as independent variables, analogous assumptions as previously yield the first-order Taylor approximation in the neighbourhood of a point $(\bar{\boldsymbol{\kappa}}, \bar{\varphi})$:

$$\boldsymbol{F}_t(\boldsymbol{\kappa}, \varphi) \approx \boldsymbol{F}_t(\bar{\boldsymbol{\kappa}}, \bar{\varphi}) + \nabla_\sigma \boldsymbol{F}_t(\bar{\boldsymbol{\kappa}}, \bar{\varphi})^{\mathrm{T}}(\boldsymbol{\kappa} - \bar{\boldsymbol{\kappa}}) + \nabla_\varphi \boldsymbol{F}_t(\bar{\boldsymbol{\kappa}}, \bar{\varphi})^{\mathrm{T}}(\varphi - \bar{\varphi}), \tag{1.30a}$$

$$M_z(\boldsymbol{\kappa}, \varphi) \approx M_z(\bar{\boldsymbol{\kappa}}, \bar{\varphi}) + \nabla_\sigma M_z(\bar{\boldsymbol{\kappa}}, \bar{\varphi})^{\mathrm{T}}(\boldsymbol{\kappa} - \bar{\boldsymbol{\kappa}}) + \nabla_\varphi M_z(\bar{\boldsymbol{\kappa}}, \bar{\varphi})^{\mathrm{T}}(\varphi - \bar{\varphi}). \tag{1.30b}$$

The practical slip stiffness matrices are accordingly defined as

$$\tilde{\mathbf{C}}_{\kappa}(\boldsymbol{\kappa}, \varphi) = \begin{bmatrix} \tilde{C}_{x\kappa_x}(\boldsymbol{\kappa}, \varphi) & \tilde{C}_{x\kappa_y}(\boldsymbol{\kappa}, \varphi) \\ \tilde{C}_{y\kappa_x}(\boldsymbol{\kappa}, \varphi) & \tilde{C}_{y\kappa_y}(\boldsymbol{\kappa}, \varphi) \end{bmatrix} \triangleq \begin{bmatrix} \dfrac{\partial F_x(\boldsymbol{\kappa}, \varphi)}{\partial \kappa_x} & \dfrac{\partial F_x(\boldsymbol{\kappa}, \varphi)}{\partial \kappa_y} \\ \dfrac{\partial F_y(\boldsymbol{\kappa}, \varphi)}{\partial \kappa_x} & \dfrac{\partial F_y(\boldsymbol{\kappa}, \varphi)}{\partial \kappa_y} \end{bmatrix} = \nabla_{\kappa} \boldsymbol{F}_t(\boldsymbol{\kappa}, \varphi)^{\mathrm{T}}, \tag{1.31a}$$

$$\tilde{\mathbf{C}}_{M\kappa}(\boldsymbol{\kappa}, \varphi) = \begin{bmatrix} \tilde{C}_{M\kappa_x}(\boldsymbol{\kappa}, \varphi) & \tilde{C}_{M\kappa_y}(\boldsymbol{\kappa}, \varphi) \end{bmatrix} \triangleq -\begin{bmatrix} \dfrac{\partial M_z(\boldsymbol{\kappa}, \varphi)}{\partial \kappa_x} & \dfrac{\partial M_z(\boldsymbol{\kappa}, \varphi)}{\partial \kappa_y} \end{bmatrix} = -\nabla_{\kappa} M_z(\boldsymbol{\kappa}, \varphi)^{\mathrm{T}}. \tag{1.31b}$$

Similarly, as for the theoretical slip stiffnesses, the entries $\tilde{C}_{x\kappa_x}$ and $\tilde{C}_{y\kappa_y}$ are called principal longitudinal and lateral slip stiffness, respectively. The terms $\tilde{C}_{x\kappa_y}$ and $\tilde{C}_{y\kappa_x}$ are instead known as cross longitudinal and lateral slip stiffnesses. Finally, the two entries $\tilde{C}_{M\kappa_x}$ and $\tilde{C}_{M\kappa_y}$ in the matrix $\tilde{\mathbf{C}}_{M\kappa}$ are the aligning longitudinal and lateral slip stiffness. All the above-mentioned stiffnesses are referred to as *practical*, to distinguish them from the theoretical ones, and *generalised* for the reasons already discussed.

Again, the conventional practical stiffnesses may be derived from Eqs. (1.31a), (1.31b) as

$$\mathbf{C}_{\kappa} = \begin{bmatrix} C_{x\kappa_x} & C_{x\kappa_y} \\ C_{y\kappa_x} & C_{y\kappa_y} \end{bmatrix} \triangleq \tilde{\mathbf{C}}_{\kappa}(\mathbf{0}, \mathbf{0}) = \nabla_{\boldsymbol{\kappa}} \boldsymbol{F}_t(\boldsymbol{\kappa}, \varphi)^{\mathrm{T}} \Big|_{\boldsymbol{\kappa}=\mathbf{0}, \varphi=0}, \tag{1.32a}$$

$$\mathbf{C}_{M\kappa} = \begin{bmatrix} C_{M\kappa_x} & C_{M\kappa_y} \end{bmatrix} \triangleq \tilde{\mathbf{C}}_{M\kappa}(\mathbf{0}, \mathbf{0}) = -\nabla_{\boldsymbol{\kappa}} M_z(\boldsymbol{\kappa}, \varphi)^{\mathrm{T}} \Big|_{\boldsymbol{\kappa}=\mathbf{0}, \varphi=0}. \tag{1.32b}$$

It is worth noticing that there are no practical stiffness matrices corresponding to $\tilde{\mathbf{C}}_{\varphi}$ and $\tilde{\mathbf{C}}_{M\varphi}$, since the practical and theoretical slips $(\boldsymbol{\kappa}, \varphi)$ and $(\boldsymbol{\sigma}, \varphi)$, respectively, only differ for the choice of the translational slip variables.

1.4.1.3 Geometrical Slip Stiffnesses

When the geometrical slips $(\kappa_x, \alpha, \gamma, \varphi)$ are used as independent variables, a first-order approximation in the neighbourhood of a point $(\bar{\kappa}_x, \bar{\alpha}, \bar{\gamma}, \bar{\varphi}_\psi)$ yields

$$\begin{aligned} \boldsymbol{F}_t(\kappa_x, \alpha, \gamma, \varphi_\psi) \approx \boldsymbol{F}_t(\bar{\kappa}_x, \bar{\alpha}, \bar{\gamma}, \bar{\varphi}_\psi) &+ \frac{\partial \boldsymbol{F}_t(\bar{\kappa}_x, \bar{\alpha}, \bar{\gamma}, \bar{\varphi}_\psi)}{\partial \kappa_x}(\kappa_x - \bar{\kappa}_x) + \frac{\partial \boldsymbol{F}_t(\bar{\kappa}_x, \bar{\alpha}, \bar{\gamma}, \bar{\varphi}_\psi)}{\partial \alpha}(\alpha - \bar{\alpha}) \\ &+ \frac{\partial \boldsymbol{F}_t(\bar{\kappa}_x, \bar{\alpha}, \bar{\gamma}, \bar{\varphi}_\psi)}{\partial \gamma}(\gamma - \bar{\gamma}) + \frac{\partial \boldsymbol{F}_t(\bar{\kappa}_x, \bar{\alpha}, \bar{\gamma}, \bar{\varphi}_\psi)}{\partial \varphi_\psi}(\varphi_\psi - \bar{\varphi}_\psi), \end{aligned} \tag{1.33a}$$

$$\begin{aligned} M_z(\kappa_x, \alpha, \gamma, \varphi_\psi) \approx M_z(\bar{\kappa}_x, \bar{\alpha}, \bar{\gamma}, \bar{\varphi}_\psi) &+ \frac{\partial M_z(\bar{\kappa}_x, \bar{\alpha}, \bar{\gamma}, \bar{\varphi}_\psi)}{\partial \kappa_x}(\kappa_x - \bar{\kappa}_x) + \frac{\partial M_z(\bar{\kappa}_x, \bar{\alpha}, \bar{\gamma}, \bar{\varphi}_\psi)}{\partial \alpha}(\alpha - \bar{\alpha}) \\ &+ \frac{\partial M_z(\bar{\kappa}_x, \bar{\alpha}, \bar{\gamma}, \bar{\varphi}_\psi)}{\partial \gamma}(\gamma - \bar{\gamma}) + \frac{\partial M_z(\bar{\kappa}_x, \bar{\alpha}, \bar{\gamma}, \bar{\varphi}_\psi)}{\partial \varphi_\psi}(\varphi_\psi - \bar{\varphi}_\psi). \end{aligned} \tag{1.33b}$$

The geometrical slip stiffnesses may be, thus, introduced as

$$\begin{bmatrix} \tilde{C}_{x\alpha}(\kappa_x, \alpha, \gamma, \varphi_\psi) \\ \tilde{C}_{y\alpha}(\kappa_x, \alpha, \gamma, \varphi_\psi) \end{bmatrix} \triangleq \begin{bmatrix} \dfrac{\partial F_x(\kappa_x, \alpha, \gamma, \varphi_\psi)}{\partial \alpha} \\ \dfrac{\partial F_y(\kappa_x, \alpha, \gamma, \varphi_\psi)}{\partial \alpha} \end{bmatrix} = \frac{\partial \boldsymbol{F}_t(\kappa_x, \alpha, \gamma, \varphi_\psi)}{\partial \alpha}, \tag{1.34a}$$

$$\begin{bmatrix} \tilde{C}_{x\gamma}\big(\kappa_x, \alpha, \gamma, \varphi_\psi\big) \\ \tilde{C}_{y\gamma}\big(\kappa_x, \alpha, \gamma, \varphi_\psi\big) \end{bmatrix} \triangleq \begin{bmatrix} \dfrac{\partial F_x\big(\kappa_x, \alpha, \gamma, \varphi_\psi\big)}{\partial \gamma} \\ \dfrac{\partial F_y\big(\kappa_x, \alpha, \gamma, \varphi_\psi\big)}{\partial \gamma} \end{bmatrix} = \frac{\partial \boldsymbol{F}_t\big(\kappa_x, \alpha, \gamma, \varphi_\psi\big)}{\partial \gamma}, \tag{1.34b}$$

$$\tilde{C}_{M\alpha}\big(\kappa_x, \alpha, \gamma, \varphi_\psi\big) \triangleq -\frac{\partial M_z\big(\kappa_x, \alpha, \gamma, \varphi_\psi\big)}{\partial \alpha}, \tag{1.34c}$$

$$\tilde{C}_{M\gamma}\big(\kappa_x, \alpha, \gamma, \varphi_\psi\big) \triangleq \frac{\partial M_z\big(\kappa_x, \alpha, \gamma, \varphi_\psi\big)}{\partial \gamma}. \tag{1.34d}$$

In this case, the representation is not as compact as those obtained for the theoretical and practical slips. The quantities $\tilde{C}_{x\alpha}$ and $\tilde{C}_{y\alpha}$ are called longitudinal and lateral cornering stiffness, whilst $\tilde{C}_{x\gamma}$ and $\tilde{C}_{y\gamma}$ longitudinal and lateral camber stiffness. The aligning cornering and camber stiffness are finally denoted by $\tilde{C}_{M\alpha}$ and $\tilde{C}_{M\gamma}$, respectively.

Once again, these stiffnesses are referred to as *geometrical* and *generalised*. Starting from Eqs. (1.34a), (1.34b), (1.34c), (1.34d), the conventional geometrical stiffnesses may be derived as

$$\begin{bmatrix} C_{x\alpha} \\ C_{y\alpha} \end{bmatrix} \triangleq \begin{bmatrix} \tilde{C}_{x\alpha}(0,0,0,0) \\ \tilde{C}_{y\alpha}(0,0,0,0) \end{bmatrix} = \begin{bmatrix} \dfrac{\partial F_x(0,0,0,0)}{\partial \alpha} \\ \dfrac{\partial F_y(0,0,0,0)}{\partial \alpha} \end{bmatrix} = \frac{\partial \boldsymbol{F}_t(0,0,0,0)}{\partial \alpha}, \tag{1.35a}$$

$$\begin{bmatrix} C_{x\gamma} \\ C_{y\gamma} \end{bmatrix} \triangleq \begin{bmatrix} \tilde{C}_{x\gamma}(0,0,0,0) \\ \tilde{C}_{y\gamma}(0,0,0,0) \end{bmatrix} = \begin{bmatrix} \dfrac{\partial F_x(0,0,0,0)}{\partial \gamma} \\ \dfrac{\partial F_y(0,0,0,0)}{\partial \gamma} \end{bmatrix} = \frac{\partial \boldsymbol{F}_t(0,0,0,0)}{\partial \gamma}, \tag{1.35b}$$

$$C_{M\alpha} \triangleq \tilde{C}_{M\alpha}(0,0,0,0) = -\frac{\partial M_z(0,0,0,0)}{\partial \alpha}, \tag{1.35c}$$

$$C_{M\gamma} \triangleq \tilde{C}_{M\gamma}(0,0,0,0) \triangleq \frac{\partial M_z(0,0,0,0)}{\partial \gamma}. \tag{1.35d}$$

1.4.2 Relationships Between Slip Stiffnesses

The theoretical and practical slip variables are related by means of the transformations given by Eqs. (1.9) and (1.10). Similarly, the geometrical slips are related to the practical ones through Eqs. (1.7), (1.11) and (1.12a), (1.12b). Therefore, it may be beneficial to understand how these transformations propagate in the relationships between the theoretical and practical slip stiffnesses, and between the practical and geometrical ones.

1.4.2.1 Relationships Between Practical and Theoretical Stiffnesses

Exploiting the transformations (1.10), the relationships between the practical and theoretical slip stiffnesses may be found using the composition rule:

$$\tilde{\mathbf{C}}_{\kappa}(\kappa, \varphi) = \nabla_{\sigma} \boldsymbol{F}_t(\sigma, \varphi)^{\mathrm{T}} \nabla_{\kappa} \sigma(\kappa)^{\mathrm{T}} \Big|_{\sigma = \frac{\kappa}{1+\kappa_x}} = \tilde{\mathbf{C}}_{\sigma}(\sigma, \varphi)\Big|_{\sigma = \frac{\kappa}{1+\kappa_x}} \begin{bmatrix} \dfrac{1}{(1+\kappa_x)^2} & 0 \\ -\dfrac{\kappa_y}{(1+\kappa_x)^2} & \dfrac{1}{1+\kappa_x} \end{bmatrix}, \tag{1.36a}$$

$$\tilde{\mathbf{C}}_{M\kappa}(\kappa, \varphi) = -\nabla_{\sigma} M_z(\sigma, \varphi)^{\mathrm{T}} \nabla_{\kappa} \sigma(\kappa)^{\mathrm{T}} \Big|_{\sigma = \frac{\kappa}{1+\kappa_x}} = \tilde{\mathbf{C}}_{M\sigma}(\sigma, \varphi)\Big|_{\sigma = \frac{\kappa}{1+\kappa_x}} \begin{bmatrix} \dfrac{1}{(1+\kappa_x)^2} & 0 \\ -\dfrac{\kappa_y}{(1+\kappa_x)^2} & \dfrac{1}{1+\kappa_x} \end{bmatrix}. \tag{1.36b}$$

From Eqs. (1.9), the inverse relationships read as follows:

$$\tilde{\mathbf{C}}_{\sigma}(\sigma, \varphi) = \nabla_{\kappa} \boldsymbol{F}_t(\kappa, \varphi)^{\mathrm{T}} \nabla_{\sigma} \kappa(\sigma)^{\mathrm{T}} \Big|_{\kappa = \frac{\sigma}{1-\sigma_x}} = \tilde{\mathbf{C}}_{\kappa}(\kappa, \varphi)\Big|_{\kappa = \frac{\sigma}{1-\sigma_x}} \begin{bmatrix} \dfrac{1}{(1-\sigma_x)^2} & 0 \\ -\dfrac{\sigma_y}{(1-\sigma_x)^2} & \dfrac{1}{1-\sigma_x} \end{bmatrix}, \tag{1.37a}$$

$$\tilde{\mathbf{C}}_{M\sigma}(\sigma, \varphi) = -\nabla_{\kappa} M_z(\kappa, \varphi)^{\mathrm{T}} \nabla_{\sigma} \kappa(\sigma)^{\mathrm{T}} \Big|_{\kappa = \frac{\sigma}{1-\sigma_x}} = \tilde{\mathbf{C}}_{M\kappa}(\kappa, \varphi)\Big|_{\kappa = \frac{\sigma}{1-\sigma_x}} \begin{bmatrix} \dfrac{1}{(1-\sigma_x)^2} & 0 \\ -\dfrac{\sigma_y}{(1-\sigma_x)^2} & \dfrac{1}{1-\sigma_x} \end{bmatrix}. \tag{1.37b}$$

As already seen, the notions of theoretical and practical slips coincide at small values of the longitudinal slips. It is very easy to verify that the latter property leads to

$$\mathbf{C}_{\kappa} \triangleq \tilde{\mathbf{C}}_{\kappa}(\mathbf{0}, \mathbf{0}) \equiv \tilde{\mathbf{C}}_{\sigma}(\mathbf{0}, \mathbf{0}) \triangleq \mathbf{C}_{\sigma}, \tag{1.38a}$$

$$\mathbf{C}_{M\kappa} \triangleq \tilde{\mathbf{C}}_{M\kappa}(\mathbf{0}, \mathbf{0}) \equiv \tilde{\mathbf{C}}_{M\sigma}(\mathbf{0}, \mathbf{0}) \triangleq \mathbf{C}_{M\sigma}. \tag{1.38b}$$

Therefore, the theoretical and practical conventional stiffnesses coincide in the origin and may be used interchangeably.

1.4.2.2 Relationships Between Geometrical and Practical Stiffnesses

From Eqs. (1.7) and (1.11), the relationships between the geometrical and practical slip stiffnesses may be found again using the composition rule. For the cornering stiffnesses,

$$\tilde{C}_{x\alpha}\big(\kappa_x, \alpha, \gamma, \varphi_\psi\big) = \frac{\partial F_x\big(\kappa_x, \alpha, \gamma, \varphi_\psi\big)}{\partial \alpha} = \frac{\partial F_x(\boldsymbol{\kappa}, \boldsymbol{\varphi})}{\partial \kappa_y}\frac{\partial \kappa_y(\alpha)}{\partial \alpha}\bigg|_{\kappa_y=\tan\alpha, \varphi_\gamma=(1-\varepsilon_\gamma)\frac{\sin\gamma}{R_r}}$$

$$= \frac{1}{\cos^2(\alpha)}\tilde{C}_{x\kappa_y}(\boldsymbol{\kappa}, \boldsymbol{\varphi})\bigg|_{\kappa_y=\tan\alpha, \varphi_\gamma=(1-\varepsilon_\gamma)\frac{\sin\gamma}{R_r}}, \tag{1.39a}$$

$$\tilde{C}_{y\alpha}\big(\kappa_x, \alpha, \gamma, \varphi_\psi\big) = \frac{\partial F_y\big(\kappa_x, \alpha, \gamma, \varphi_\psi\big)}{\partial \alpha} = \frac{\partial F_y(\boldsymbol{\kappa}, \boldsymbol{\varphi})}{\partial \kappa_y}\frac{\partial \kappa_y(\alpha)}{\partial \alpha}\bigg|_{\kappa_y=\tan\alpha, \varphi_\gamma=(1-\varepsilon_\gamma)\frac{\sin\gamma}{R_r}}$$

$$= \frac{1}{\cos^2(\alpha)}\tilde{C}_{y\kappa_y}(\boldsymbol{\kappa}, \boldsymbol{\varphi})\bigg|_{\kappa_y=\tan\alpha, \varphi_\gamma=(1-\varepsilon_\gamma)\frac{\sin\gamma}{R_r}}, \tag{1.39b}$$

$$\tilde{C}_{M\alpha}\big(\kappa_x, \alpha, \gamma, \varphi_\psi\big) = -\frac{\partial M_z\big(\kappa_x, \alpha, \gamma, \varphi_\psi\big)}{\partial \alpha} = -\frac{\partial M_z(\boldsymbol{\kappa}, \boldsymbol{\varphi})}{\partial \kappa_y}\frac{\partial \kappa_y(\alpha)}{\partial \alpha}\bigg|_{\kappa_y=\tan\alpha, \varphi_\gamma=(1-\varepsilon_\gamma)\frac{\sin\gamma}{R_r}}$$

$$= \frac{1}{\cos^2(\alpha)}\tilde{C}_{M\kappa_y}(\boldsymbol{\kappa}, \boldsymbol{\varphi})\bigg|_{\kappa_y=\tan\alpha, \varphi_\gamma=(1-\varepsilon_\gamma)\frac{\sin\gamma}{R_r}}. \tag{1.39c}$$

Analogously, for the camber stiffness,

$$\tilde{C}_{x\gamma}\big(\kappa_x, \alpha, \gamma, \varphi_\psi\big) = \frac{\partial F_x\big(\kappa_x, \alpha, \gamma, \varphi_\psi\big)}{\partial \gamma} = \frac{\partial F_x(\boldsymbol{\kappa}, \boldsymbol{\varphi})}{\partial \varphi_\gamma}\frac{\partial \varphi_\gamma(\gamma)}{\partial \gamma}\bigg|_{\kappa_y=\tan\alpha, \varphi_\gamma=(1-\varepsilon_\gamma)\frac{\sin\gamma}{R_r}}$$

$$= \big(1-\varepsilon_\gamma\big)\frac{\cos\gamma}{R_r}\tilde{C}_{x\varphi_\gamma}(\boldsymbol{\kappa}, \boldsymbol{\varphi})\bigg|_{\kappa_y=\tan\alpha, \varphi_\gamma=(1-\varepsilon_\gamma)\frac{\sin\gamma}{R_r}}, \tag{1.40a}$$

$$\tilde{C}_{y\gamma}\big(\kappa_x, \alpha, \gamma, \varphi_\psi\big) = \frac{\partial F_y\big(\kappa_x, \alpha, \gamma, \varphi_\psi\big)}{\partial \gamma} = \frac{\partial F_y(\boldsymbol{\kappa}, \boldsymbol{\varphi})}{\partial \varphi_\gamma}\frac{\partial \varphi_\gamma(\gamma)}{\partial \gamma}\bigg|_{\kappa_y=\tan\alpha, \varphi_\gamma=(1-\varepsilon_\gamma)\frac{\sin\gamma}{R_r}}$$

$$= \big(1-\varepsilon_\gamma\big)\frac{\cos\gamma}{R_r}\tilde{C}_{y\varphi_\gamma}(\boldsymbol{\kappa}, \boldsymbol{\varphi})\bigg|_{\kappa_y=\tan\alpha, \varphi_\gamma=(1-\varepsilon_\gamma)\frac{\sin\gamma}{R_r}}, \tag{1.40b}$$

$$\tilde{C}_{M\gamma}\big(\kappa_x, \alpha, \gamma, \varphi_\psi\big) = \frac{\partial M_z\big(\kappa_x, \alpha, \gamma, \varphi_\psi\big)}{\partial \gamma} = \frac{\partial M_z(\boldsymbol{\kappa}, \boldsymbol{\varphi})}{\partial \varphi_\gamma} \frac{\partial \varphi_\gamma(\gamma)}{\partial \gamma}\bigg|_{\kappa_y=\tan\alpha, \varphi_\gamma=(1-\varepsilon_\gamma)\frac{\sin\gamma}{R_r}}$$

$$= \big(1-\varepsilon_\gamma\big)\frac{\cos\gamma}{R_r}\tilde{C}_{M\varphi_\gamma}(\boldsymbol{\kappa}, \boldsymbol{\varphi})\bigg|_{\kappa_y=\tan\alpha, \varphi_\gamma=(1-\varepsilon_\gamma)\frac{\sin\gamma}{R_r}}. \qquad (1.40c)$$

Using Eqs. (1.12a), (1.12b), the inverse relationships are, respectively, given by

$$\tilde{C}_{x\kappa_y}(\boldsymbol{\kappa}, \boldsymbol{\varphi}) = \frac{\partial F_x(\boldsymbol{\kappa}, \boldsymbol{\varphi})}{\partial \kappa_y} = \frac{\partial F_x\big(\kappa_x, \alpha, \gamma, \varphi_\psi\big)}{\partial \alpha}\frac{\partial \alpha\big(\kappa_y\big)}{\partial \kappa_y}\bigg|_{\alpha=\arctan\kappa_y, \gamma=\arcsin\left(\frac{\varphi_\gamma R_r}{1-\varepsilon_\gamma}\right)}$$

$$= \frac{1}{1+\kappa_y^2}\tilde{C}_{x\alpha}\big(\kappa_x, \alpha, \gamma, \varphi_\psi\big)\bigg|_{\alpha=\arctan\kappa_y, \gamma=\arcsin\left(\frac{\varphi_\gamma R_r}{1-\varepsilon_\gamma}\right)}, \qquad (1.41a)$$

$$\tilde{C}_{y\kappa_y}(\boldsymbol{\kappa}, \boldsymbol{\varphi}) = \frac{\partial F_y(\boldsymbol{\kappa}, \boldsymbol{\varphi})}{\partial \kappa_y} = \frac{\partial F_y\big(\kappa_x, \alpha, \gamma, \varphi_\psi\big)}{\partial \alpha}\frac{\partial \alpha\big(\kappa_y\big)}{\partial \kappa_y}\bigg|_{\alpha=\arctan\kappa_y, \gamma=\arcsin\left(\frac{\varphi_\gamma R_r}{1-\varepsilon_\gamma}\right)}$$

$$= \frac{1}{1+\kappa_y^2}\tilde{C}_{y\alpha}\big(\kappa_x, \alpha, \gamma, \varphi_\psi\big)\bigg|_{\alpha=\arctan\kappa_y, \gamma=\arcsin\left(\frac{\varphi_\gamma R_r}{1-\varepsilon_\gamma}\right)}, \qquad (1.41b)$$

$$\tilde{C}_{M\kappa_y}(\boldsymbol{\kappa}, \boldsymbol{\varphi}) = -\frac{\partial M_z(\boldsymbol{\kappa}, \boldsymbol{\varphi})}{\partial \kappa_y} = -\frac{\partial M_z\big(\kappa_x, \alpha, \gamma, \varphi_\psi\big)}{\partial \alpha}\frac{\partial \alpha\big(\kappa_y\big)}{\partial \kappa_y}\bigg|_{\alpha=\arctan\kappa_y, \gamma=\arcsin\left(\frac{\varphi_\gamma R_r}{1-\varepsilon_\gamma}\right)}$$

$$= \frac{1}{1+\kappa_y^2}\tilde{C}_{M\alpha}\big(\kappa_x, \alpha, \gamma, \varphi_\psi\big)\bigg|_{\alpha=\arctan\kappa_y, \gamma=\arcsin\left(\frac{\varphi_\gamma R_r}{1-\varepsilon_\gamma}\right)}, \qquad (1.41c)$$

and

$$\tilde{C}_{x\varphi_\gamma}(\boldsymbol{\kappa}, \boldsymbol{\varphi}) = \frac{\partial F_x(\boldsymbol{\kappa}, \boldsymbol{\varphi})}{\partial \varphi_\gamma} = \frac{\partial F_x\big(\kappa_x, \alpha, \gamma, \varphi_\psi\big)}{\partial \gamma}\frac{\partial \gamma\big(\varphi_\gamma\big)}{\partial \varphi_\gamma}\bigg|_{\alpha=\arctan\kappa_y, \gamma=\arcsin\left(\frac{\varphi_\gamma R_r}{1-\varepsilon_\gamma}\right)}$$

$$= \frac{R_r}{\sqrt{\big(1-\varepsilon_\gamma\big)^2 - \varphi_\gamma^2 R_r^2}}\tilde{C}_{x\gamma}\big(\kappa_x, \alpha, \gamma, \varphi_\psi\big)\bigg|_{\alpha=\arctan\kappa_y, \gamma=\arcsin\left(\frac{\varphi_\gamma R_r}{1-\varepsilon_\gamma}\right)}, \qquad (1.42a)$$

$$\tilde{C}_{y\varphi_\gamma}(\boldsymbol{\kappa}, \boldsymbol{\varphi}) = \frac{\partial F_y(\boldsymbol{\kappa}, \boldsymbol{\varphi})}{\partial \varphi_\gamma} = \frac{\partial F_y\big(\kappa_x, \alpha, \gamma, \varphi_\psi\big)}{\partial \gamma} \frac{\partial \gamma(\varphi_\gamma)}{\partial \varphi_\gamma}\Bigg|_{\alpha=\arctan \kappa_y, \gamma=\arcsin\left(\frac{\varphi_\gamma R_r}{1-\varepsilon_\gamma}\right)}$$

$$= \frac{R_r}{\sqrt{\big(1-\varepsilon_\gamma\big)^2 - \varphi_\gamma^2 R_r^2}} \tilde{C}_{y\gamma}\big(\kappa_x, \alpha, \gamma, \varphi_\psi\big)\Bigg|_{\alpha=\arctan \kappa_y, \gamma=\arcsin\left(\frac{\varphi_\gamma R_r}{1-\varepsilon_\gamma}\right)}, \tag{1.42b}$$

$$\tilde{C}_{M\varphi_\gamma}(\boldsymbol{\kappa}, \boldsymbol{\varphi}) = \frac{\partial M_z(\boldsymbol{\kappa}, \boldsymbol{\varphi})}{\partial \varphi_\gamma} = \frac{\partial M_z\big(\kappa_x, \alpha, \gamma, \varphi_\psi\big)}{\partial \gamma} \frac{\partial \gamma(\varphi_\gamma)}{\partial \varphi_\gamma}\Bigg|_{\alpha=\arctan \kappa_y, \gamma=\arcsin\left(\frac{\varphi_\gamma R_r}{1-\varepsilon_\gamma}\right)}$$

$$= \frac{R_r}{\sqrt{\big(1-\varepsilon_\gamma\big)^2 - \varphi_\gamma^2 R_r^2}} \tilde{C}_{M\gamma}\big(\kappa_x, \alpha, \gamma, \varphi_\psi\big)\Bigg|_{\alpha=\arctan \kappa_y, \gamma=\arcsin\left(\frac{\varphi_\gamma R_r}{1-\varepsilon_\gamma}\right)}. \tag{1.42c}$$

Also, in this case, the notions of cornering stiffnesses and the practical lateral stiffnesses coincide in the origin. In vector notation,

$$\begin{bmatrix} C_{x\alpha} \\ C_{y\alpha} \\ C_{M\alpha} \end{bmatrix} \triangleq \begin{bmatrix} \tilde{C}_{x\alpha}(0,0,0,0) \\ \tilde{C}_{y\alpha}(0,0,0,0) \\ \tilde{C}_{M\alpha}(0,0,0,0) \end{bmatrix} \equiv \begin{bmatrix} \tilde{C}_{x\kappa_y}(\mathbf{0},\mathbf{0}) \\ \tilde{C}_{y\kappa_y}(\mathbf{0},\mathbf{0}) \\ \tilde{C}_{M\kappa_y}(\mathbf{0},\mathbf{0}) \end{bmatrix} \triangleq \begin{bmatrix} C_{x\kappa_y} \\ C_{y\kappa_y} \\ C_{M\kappa_y} \end{bmatrix}. \tag{1.43}$$

On the other hand, the relationships between the conventional camber stiffnesses and the camber spin stiffnesses are given by

$$\begin{bmatrix} C_{x\gamma} \\ C_{y\gamma} \\ C_{M\gamma} \end{bmatrix} \triangleq \begin{bmatrix} \tilde{C}_{x\gamma}(0,0,0,0) \\ \tilde{C}_{y\gamma}(0,0,0,0) \\ \tilde{C}_{M\gamma}(0,0,0,0) \end{bmatrix} \equiv \frac{1}{R_r}(1-\varepsilon_\gamma) \begin{bmatrix} \tilde{C}_{x\varphi_\gamma}(\mathbf{0},\mathbf{0}) \\ \tilde{C}_{y\varphi_\gamma}(\mathbf{0},\mathbf{0}) \\ \tilde{C}_{M\varphi_\gamma}(\mathbf{0},\mathbf{0}) \end{bmatrix} \triangleq \frac{1}{R_r}(1-\varepsilon_\gamma) \begin{bmatrix} C_{x\varphi_\gamma} \\ C_{y\varphi_\gamma} \\ C_{M\varphi_\gamma} \end{bmatrix}. \tag{1.44}$$

In Eq. (1.44), the additional term $(1-\varepsilon_\gamma)/R_r$ appears because, whilst the camber angle γ is a nondimensional quantity, the camber spin φ_γ has the dimension of a curvature. The coefficient $(1-\varepsilon_\gamma)/R_r$ also coincides with the limit camber spin (corresponding to a camber angle of 90°).

1.4.3 Slip-Tyre Steady-State Relationships

As briefly mentioned in Sect. 1.4.1, Eqs. (1.22a), (1.22b), (1.23a), (1.23b) and (1.24a), (1.24b) represent the most common way—and also the natural one—to describe the tyre-road interaction in steady-state conditions. However, in view of some applications, it might be beneficial to regard the tyre forces and moment as independent variables and treat the slips as static nonlinearities [9].

Often, these relationships must be derived inverting Eqs. (1.22a), (1.22b), (1.23a), (1.23b) and (1.24a), (1.24b), which are locally diffeomorphic. This may be done under the assumption of small spin slips, so as to condense in the total spin slip φ the individual contributions of the geometrical and camber spin. In this case, the slips may be represented explicitly as a function of the triplet $(\boldsymbol{F}_t, M_z)$ as

$$\boldsymbol{\sigma} = \boldsymbol{\sigma}(\boldsymbol{F}_t, M_z), \tag{1.45a}$$

$$\varphi = \varphi(\boldsymbol{F}_t, M_z), \tag{1.45b}$$

or

$$\boldsymbol{\kappa} = \boldsymbol{\kappa}(\boldsymbol{F}_t, M_z), \tag{1.46a}$$

$$\varphi = \varphi(\boldsymbol{F}_t, M_z), \tag{1.46b}$$

or

$$\kappa_x = \kappa_x(\boldsymbol{F}_t, M_z), \tag{1.47a}$$

$$\alpha = \alpha(\boldsymbol{F}_t, M_z), \tag{1.47b}$$

$$\gamma = \gamma(\boldsymbol{F}_t, M_z). \tag{1.47c}$$

In this context, it should be emphasised that using the geometrical slip $(\kappa_x, \alpha, \gamma)$ as dependent variables automatically implies that the effect of the turn spin φ_ψ is neglected in the computation of the total one φ.

In particular, starting from the explicit relationships (1.22a), (1.22b), (1.23a), (1.23b) and (1.24a), (1.24b), local representations of the type (1.45a), (1.45b), (1.46a), (1.46b) and (1.47a), (1.47b), (1.47c) may be certainly found in the neighbourhood of a point $(\bar{\boldsymbol{\sigma}}, \bar{\varphi})$, $(\bar{\boldsymbol{\kappa}}, \bar{\varphi})$ or $(\bar{\kappa}_x, \bar{\alpha}, \bar{\gamma})$ if the following conditions are, respectively, fulfilled:

$$\det \begin{bmatrix} \tilde{\mathbf{C}}_{\sigma}(\bar{\boldsymbol{\sigma}}, \bar{\varphi}) & \tilde{\mathbf{C}}_{\varphi}(\bar{\boldsymbol{\sigma}}, \bar{\varphi}) \\ \tilde{\mathbf{C}}_{M\sigma}(\bar{\boldsymbol{\sigma}}, \bar{\varphi}) & \tilde{C}_{M\varphi}(\bar{\boldsymbol{\sigma}}, \bar{\varphi}) \end{bmatrix} \neq 0, \tag{1.48a}$$

$$\det \begin{bmatrix} \tilde{\mathbf{C}}_{\kappa}(\bar{\boldsymbol{\kappa}}, \bar{\varphi}) & \tilde{\mathbf{C}}_{\varphi}(\bar{\boldsymbol{\kappa}}, \bar{\varphi}) \\ \tilde{\mathbf{C}}_{M\kappa}(\bar{\boldsymbol{\kappa}}, \bar{\varphi}) & \tilde{C}_{M\varphi}(\bar{\boldsymbol{\kappa}}, \bar{\varphi}) \end{bmatrix} \neq 0, \tag{1.48b}$$

$$\det \begin{bmatrix} \tilde{C}_{x\kappa_x}(\bar{\kappa}_x, \bar{\alpha}, \bar{\gamma}) & \tilde{C}_{x\alpha}(\bar{\kappa}_x, \bar{\alpha}, \bar{\gamma}) & \tilde{C}_{x\gamma}(\bar{\kappa}_x, \bar{\alpha}, \bar{\gamma}) \\ \tilde{C}_{y\kappa_x}(\bar{\kappa}_x, \bar{\alpha}, \bar{\gamma}) & \tilde{C}_{y\alpha}(\bar{\kappa}_x, \bar{\alpha}, \bar{\gamma}) & \tilde{C}_{y\gamma}(\bar{\kappa}_x, \bar{\alpha}, \bar{\gamma}) \\ \tilde{C}_{M\kappa_x}(\bar{\kappa}_x, \bar{\alpha}, \bar{\gamma}) & \tilde{C}_{M\alpha}(\bar{\kappa}_x, \bar{\alpha}, \bar{\gamma}) & \tilde{C}_{M\gamma}(\bar{\kappa}_x, \bar{\alpha}, \bar{\gamma}) \end{bmatrix} \neq 0. \tag{1.48c}$$

The requirements in Eqs. (1.48a), (1.48b), (1.48c) satisfy the assumptions of the Inverse Function Theorem. Furthermore, it may be easily shown that

$$\det\begin{bmatrix} \tilde{\mathbf{C}}_\sigma & \tilde{\mathbf{C}}_\varphi \\ \tilde{\mathbf{C}}_{M\sigma} & \tilde{C}_{M\varphi} \end{bmatrix} = \det\left(\tilde{\mathbf{C}}_\sigma - \frac{1}{\tilde{C}_{M\varphi}}\tilde{\mathbf{C}}_\varphi\tilde{\mathbf{C}}_{M\sigma}\right)$$
$$\equiv \det\left(\tilde{\mathbf{C}}_\kappa - \frac{1}{\tilde{C}_{M\varphi}}\tilde{\mathbf{C}}_\varphi\tilde{\mathbf{C}}_{M\kappa}\right)\det\nabla_{\boldsymbol{\sigma}}\boldsymbol{\kappa}^{\mathrm{T}} = \frac{1}{(1-\sigma_x)^3}\det\begin{bmatrix} \tilde{\mathbf{C}}_\kappa & \tilde{\mathbf{C}}_\varphi \\ \tilde{\mathbf{C}}_{M\kappa} & \tilde{C}_{M\varphi} \end{bmatrix}. \tag{1.49}$$

Therefore,

$$\det\begin{bmatrix} \tilde{\mathbf{C}}_\sigma & \tilde{\mathbf{C}}_\varphi \\ \tilde{\mathbf{C}}_{M\sigma} & \tilde{C}_{M\varphi} \end{bmatrix} = 0 \iff \det\begin{bmatrix} \tilde{\mathbf{C}}_\kappa & \tilde{\mathbf{C}}_\varphi \\ \tilde{\mathbf{C}}_{M\kappa} & \tilde{C}_{M\varphi} \end{bmatrix} = 0. \tag{1.50}$$

Analogously, for any value of $|\gamma| < \pi/2$,

$$\det\begin{bmatrix} \tilde{\mathbf{C}}_\kappa & \tilde{\mathbf{C}}_\varphi \\ \tilde{\mathbf{C}}_{M\kappa} & \tilde{C}_{M\varphi} \end{bmatrix} = \det\left(\tilde{\mathbf{C}}_\kappa - \frac{1}{\tilde{C}_{M\varphi}}\tilde{\mathbf{C}}_\varphi\tilde{\mathbf{C}}_{M\kappa}\right)$$
$$\equiv \det\left(\begin{bmatrix} \tilde{C}_{x\kappa} & \tilde{C}_{x\alpha} \\ \tilde{C}_{y\kappa} & \tilde{C}_{y\alpha} \end{bmatrix} - \frac{1}{\tilde{C}_{M\gamma}}\begin{bmatrix} \tilde{C}_{x\gamma} \\ \tilde{C}_{y\gamma} \end{bmatrix}\begin{bmatrix} \tilde{C}_{M\kappa_x} & \tilde{C}_{M\alpha} \end{bmatrix}\right)\det\begin{bmatrix} 1 & 0 \\ 0 & 1+\kappa_y^2 \end{bmatrix}$$
$$= \left(1+\kappa_y^2\right)\det\begin{bmatrix} \tilde{C}_{x\kappa_x} & \tilde{C}_{x\alpha} & \tilde{C}_{x\gamma} \\ \tilde{C}_{y\kappa_x} & \tilde{C}_{y\alpha} & \tilde{C}_{y\gamma} \\ \tilde{C}_{M\kappa_x} & \tilde{C}_{M\alpha} & \tilde{C}_{M\gamma} \end{bmatrix}. \tag{1.51}$$

Therefore,

$$\det\begin{bmatrix} \tilde{\mathbf{C}}_\kappa & \tilde{\mathbf{C}}_\varphi \\ \tilde{\mathbf{C}}_{M\kappa} & \tilde{C}_{M\varphi} \end{bmatrix} = 0 \iff \det\begin{bmatrix} \tilde{C}_{x\kappa_x} & \tilde{C}_{x\alpha} & \tilde{C}_{x\gamma} \\ \tilde{C}_{y\kappa_x} & \tilde{C}_{y\alpha} & \tilde{C}_{y\gamma} \\ \tilde{C}_{M\kappa_x} & \tilde{C}_{M\alpha} & \tilde{C}_{M\gamma} \end{bmatrix} = 0. \tag{1.52}$$

Owing to Eqs. (1.50) and (1.52), it may be concluded that the three requirements given by Eqs. (1.48a), (1.48b), (1.48c) are equivalent, provided that $\sigma_x < 1$ and $s|\alpha|, |\gamma| < \pi/2$. In normal operating conditions, the preceding inequalities are always fulfilled.

References

1. Pacejka HB (2012) Tire and vehicle dynamics, 3rd edn. Elsevier/BH, Amsterdam
2. Lugner P (2019) Vehicle Dynamics of Modern Passenger Cars. Springer International Publishing Copyright Holder
3. Guiggiani M (2018) The science of vehicle dynamics, 2nd edn. Springer International, Cham(Switzerland)
4. Svendenius J (2007) Tire modelling and friction estimation [dissertation]. Lund
5. Pacejka HB (2005) Spin: camber and turning. Veh Syst Dyn 43(1):3–17. https://doi.org/10.1080/00423110500140013

6. Mavros G (2019) A thermo-frictional tyre model including the effect of flash temperature. Veh Syst Dyn 57(5):721–751. https://doi.org/10.1080/00423114.2018.1484147
7. Sakhnevych A (2021) Multiphysical MF-based tyre modelling and parametrisation for vehicle setup and control strategies optimisation. Veh Syst Dyn. https://doi.org/10.1080/00423114.2021.1977833
8. Romano L, Bruzelius F, Jacobson B. Brush tyre models for large camber angles and steering speeds. Veh Syst Dyn. https://doi.org/10.1080/00423114.2020.1854320
9. Bruzelius F, Hjort M, Svendenius J (2014) Validation of a basic combined-slip tyre model for use in friction estimation applications. In: Proceedings of the institution of mechanical engineers. Part D: J Automob Eng 228(13):1622–1629. https://doi.org/10.1177/0954407013511797

Chapter 2
Tyre-Road Contact Mechanics Equations

Abstract The brush theory constitutes the simplest, full-physical approach to model the tyre-road interaction. It describes the tyre-wheel system as a rigid body equipped with bristles, which undergo tangential deformations in their attempt to stick to the ground. This chapter introduces the governing equations of the brush models. These may be divided into four separate sets: the friction model, the constitutive relationships, the tyre-road kinematic relationships and the equilibrium equations. The fundamental assumptions behind the proposed formulation are outlined, and the boundary and initial conditions are stated in a general manner.

In the brush models, the interaction between the tyre and the road is described in a reference frame $(O; x, y, z)$ with unit vectors $(\hat{\boldsymbol{e}}_x, \hat{\boldsymbol{e}}_y, \hat{\boldsymbol{e}}_z)$. The origin O is contact-fixed, that is attached to the tyre contact patch. When the tyre carcass is undeformed, $O \equiv C$. Furthermore, if the contact patch is symmetric, then both O and C coincide with its centre. As illustrated in Fig. 2.1, the axes are oriented according to the convention adopted by Pacejka [1]: the x-axis is directed towards the longitudinal direction of motion, the z-axis points downward and the y-axis lies on the road surface and is oriented so that the coordinate system is right-handed. The road is modelled as a perfectly homogeneous, isotropic flat surface, without any irregularity. When the effect of a compliant carcass is disregarded, the tyre-wheel system is considered rigid once the normal contact has occurred and equipped with bristles that only undergo tangential deformations on the road plane $\Pi = \{\boldsymbol{x} \in \mathbb{R}^3 \mid z = 0\}$. Tangential and normal vectors to the plane Π are denoted by $\boldsymbol{t} = x\hat{\boldsymbol{e}}_x + y\hat{\boldsymbol{e}}_y$ and $\boldsymbol{n} = z\hat{\boldsymbol{e}}_z$, respectively. The contact patch collects all the points $\boldsymbol{x} \in \Pi$ where the tyre and the road come in contact. It is defined mathematically as a possibly time-varying compact set $\mathscr{P} = \mathscr{P}(t)$, whose interior and boundary are denoted by $\mathring{\mathscr{P}} = \mathring{\mathscr{P}}(t)$ and $\partial\mathscr{P} = \partial\mathscr{P}(t)$, respectively, with initial conditions $\mathscr{P}_0 \triangleq \mathscr{P}(0)$, $\partial\mathscr{P}_0 \triangleq \partial\mathscr{P}(0)$ and $\mathring{\mathscr{P}}_0 \triangleq \mathring{\mathscr{P}}(0)$.

During the rolling of the tyre, a generic quantity may evolve over time $t \in \mathbb{R}_{\geq 0}$ or, equivalently, over a peculiar distance $s \in \mathbb{R}_{\geq 0}$, called *travelled distance* and defined by

$$s = \int_0^t V_{\mathrm{r}}(t')\,\mathrm{d}t' \implies \mathrm{d}s = V_{\mathrm{r}}(t)\,\mathrm{d}t, \tag{2.1}$$

L. Romano, *Advanced Brush Tyre Modelling*,
SpringerBriefs in Applied Sciences and Technology,
https://doi.org/10.1007/978-3-030-98435-9_2

where $V_r = \Omega R_r$ has been introduced in Chap. 1. In this context, the definition of rolling radius that already includes the offset $c_r(\gamma)$ is appropriate, because V_r actually represents the velocity of the contact patch centre (more precisely, of the origin O) in the Eulerian approach. Other choices for the origin O are also possible, provided that it is contact-fixed.

The mapping $t \to s$ is one-to-one, since Ω and V_r are both assumed greater than zero. This makes it possible to use the time t or the travelled distance s interchangeably. In particular, replacing the time variable with the travelled distance allows formulating the equations of the tyre-road kinematics so that they are independent of the rolling speed of the tyre. In this case, the response of the tyre is fully determined by the slip inputs, whereas the velocities do not play any role. Therefore, in the following, all the equations will be stated using s as an independent variable. It should be noticed that other velocities also appear in the nondimensional form and are indicated using the notation $\bar{(\cdot)}$; the corresponding dimensional velocities may be obtained by multiplying by V_r.

In particular, the governing equations of the brush theory are formulated according to the Eulerian approach. Thus, a nondimensional velocity field $\mathrm{d}\boldsymbol{x}/\mathrm{d}s = \bar{\boldsymbol{v}}_t(\boldsymbol{x}, s) = \bar{v}_x(\boldsymbol{x}, s)\hat{\boldsymbol{e}}_x + \bar{v}_y(\boldsymbol{x}, s)\hat{\boldsymbol{e}}_y$ is associated to each point $\boldsymbol{x} \in \mathscr{P}$ that describes the nondimensional tangential velocity of the root of the bristles, which is attached to the contact patch. On the other hand, the velocity of the tip of the bristles contacting the ground is called *micro-sliding velocity*. In nondimensional form, it is denoted by $\bar{\boldsymbol{v}}_s(\boldsymbol{x}, s) = \bar{v}_{sx}(\boldsymbol{x}, s)\hat{\boldsymbol{e}}_x + \bar{v}_{sy}(\boldsymbol{x}, s)\hat{\boldsymbol{e}}_y$.

To adhere to the road, the tips of the bristles try to stick to the ground, and, as a consequence, they undergo a tangential deflection (deformation), described by $\boldsymbol{u}_t(\boldsymbol{x}, s) = u_x(\boldsymbol{x}, s)\hat{\boldsymbol{e}}_x + u_y(\boldsymbol{x}, s)\hat{\boldsymbol{e}}_y$. This quantity represents the relative position between the root of a bristle and its tip. Each bristle is also subjected to forces per unit of area, which may vary both in space and time. In particular, both normal and the tangential stresses $q_z(\boldsymbol{x}, s)$ and $\boldsymbol{q}_t(\boldsymbol{x}, s) = q_x(\boldsymbol{x}, t)\hat{\boldsymbol{e}}_x + q_y(\boldsymbol{x}, s)\hat{\boldsymbol{e}}_y$, respectively, act upon the bristles. The stresses $q_z(\boldsymbol{x}, s)$ are also referred to as *vertical pressure*, whilst $\boldsymbol{q}_t(\boldsymbol{x}, s)$ are often called *shear stresses*. A schematic of the tyre equipped with bristles is illustrated in Fig. 2.1.

2.1 Friction Model

In the brush theory, it is assumed that the tangential problem is decoupled from the vertical one; this implies that the vertical pressure distribution is not influenced by the shear stresses acting inside the contact patch.[1] Owing to these premises, the fundamental equations governing the tyre-road contact mechanics may be formulated adopting a simple Coulomb friction model as follows [1]:

---

[1] There is no well-established theory for the case in which the distribution $q_z(\boldsymbol{x}, t)$ is affected by friction-induced phenomena. See [2, 3].

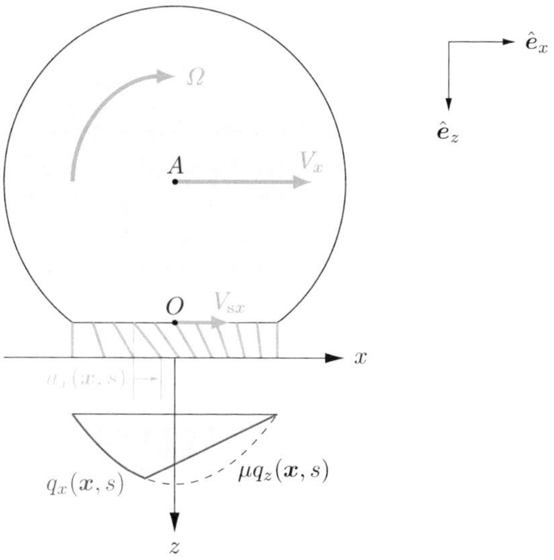

Fig. 2.1 Schematic of the tyre-wheel systems equipped with bristles. The bristles undergo a deformation in the road plane and are subjected to tangential shear stresses. The magnitude of these is limited by the available friction

$$\bar{\boldsymbol{v}}_{\mathrm{s}}(\boldsymbol{x}, s) = \boldsymbol{0} \implies q_t(\boldsymbol{x}, s) \leq \mu q_z(\boldsymbol{x}, s), \tag{2.2a}$$

$$\bar{\boldsymbol{v}}_{\mathrm{s}}(\boldsymbol{x}, s) \neq \boldsymbol{0} \iff \boldsymbol{q}_t(\boldsymbol{x}, s) = \mu q_z(\boldsymbol{x})\hat{\boldsymbol{s}}_t(\boldsymbol{x}, s), \tag{2.2b}$$

in which *sliding direction* $\hat{\boldsymbol{s}}_t(\boldsymbol{x}, s)$ is written as

$$\hat{\boldsymbol{s}}_t(\boldsymbol{x}, s) \triangleq -\frac{\bar{\boldsymbol{v}}_{\mathrm{s}}(\boldsymbol{x}, s)}{\bar{v}_{\mathrm{s}}(\boldsymbol{x}, s)}. \tag{2.3}$$

In Eqs. (2.2a), (2.2b), $q_t(\boldsymbol{x}, s) \triangleq \left\| \boldsymbol{q}_t(\boldsymbol{x}, s) \right\|$, where $\|\cdot\|$ is the Euclidean norm, and μ is the friction coefficient. Analogously, $\bar{v}_{\mathrm{s}}(\boldsymbol{x}, s) \triangleq \left\| \bar{\boldsymbol{v}}_{\mathrm{s}}(\boldsymbol{x}, s) \right\|$. In general, both quantities may depend explicitly upon the vector position $\boldsymbol{x}$ inside the contact patch or the nondimensional sliding velocity $\bar{\boldsymbol{v}}_{\mathrm{s}}(\boldsymbol{x}, s)$. However, to keep the analysis within acceptable levels, they are assumed constant in this book. More refined theories may be found in [4–17].

Equation (2.2a) defines the condition that allows a bristle to stick to the ground: the total shear stress acting upon it must not exceed the friction limit $\mu q_z(\boldsymbol{x}, s)$. When this happens, even deforming, the bristle cannot compensate for the difference between the velocity of its root and that of the road and starts suddenly to slide. As a consequence, the nondimensional tangential velocity of the tip increases abruptly in magnitude, that is $\bar{\boldsymbol{v}}_{\mathrm{s}}(\boldsymbol{x}, s) \neq \boldsymbol{0}$. The dynamics of the bristle is governed by different phenomena depending on whether it adheres to the road or slide. Since these phenomena take place in different portions of the contact patch, it is customary to partition it into an adhesion region $\mathscr{P}^{(\mathrm{a})}$ and a sliding one $\mathscr{P}^{(\mathrm{s})}$. These may be mathematically defined as

$$\mathscr{P}^{(\mathrm{a})} \triangleq \left\{ \boldsymbol{x} \in \mathscr{P} \mid \text{Eq. 2.2a holds} \right\}, \tag{2.4a}$$

$$\mathscr{P}^{(\mathrm{s})} \triangleq \left\{ \boldsymbol{x} \in \mathscr{P} \mid \text{Eq. 2.2b holds} \right\}, \tag{2.4b}$$

clearly with $\mathscr{P} = \mathscr{P}^{(\mathrm{a})} \cup \mathscr{P}^{(\mathrm{s})}$. Keeping the notation used in Eqs. (2.4a), (2.4b), a generic quantity may be denoted by $(\cdot)^{(\mathrm{a})}(\boldsymbol{x})$ if $\boldsymbol{x} \in \mathscr{P}^{(\mathrm{a})}$ and by $(\cdot)^{(\mathrm{s})}(\boldsymbol{x})$ if $\boldsymbol{x} \in \mathscr{P}^{(\mathrm{s})}$.

To solve the above Eqs. (2.2a), (2.2b), three other sets of relationships are needed: the *constitutive equations*, the *tyre-road kinematic equations* and the *equilibrium equations*. Specifically, the first set prescribes the relationships between the shear stress acting upon the bristle and its deflection, and those between the tangential forces acting on the tyre and the displacement of the tyre carcass, respectively. The tyre-road kinematic equations provide the relationship between the sliding velocity and the deflection of the bristles. Finally, the equilibrium equations provide the integral relationships that allow expressing the tyre characteristics as a function of the shear stress acting upon the bristles. It should be noticed that the tyre forces only enter the kinematic equations if the deformation of the tyre carcass is taken into account.

2.2 Constitutive Relationships

The first set of constitutive equations relates the deflection of a bristle to the shear stress acting on its tip. The second set relates the deformation of the tyre carcass and the tangential forces acting at the contact patch.

2.2.1 Constitutive Relationships for the Tyre Tread

These equations establish the relationships between the local shear stress $\boldsymbol{q}_t(\boldsymbol{x}, s)$ acting inside the contact patch and the tangential deflection of the bristle $\boldsymbol{u}_t(\boldsymbol{x}, s)$. In spite of the viscoelastic nature of the tyre, for the sake of simplicity, it is customary in the literature to assume linear elasticity [4], that is a constitutive relation of the type

$$\boldsymbol{q}_t(\boldsymbol{x}, s) = \mathbf{K}_t \boldsymbol{u}_t(\boldsymbol{x}, s), \tag{2.5}$$

where the tangential stiffness matrix $\mathbf{K}_t$ reads

$$\mathbf{K}_t = \begin{bmatrix} k_{xx} & k_{xy} \\ k_{yx} & k_{yy} \end{bmatrix}, \tag{2.6}$$

and it is assumed to be positive definite (often even diagonal).

2.2.2 Constitutive Relationships for the Tyre Carcass

The tyre carcass is usually modelled as a linear spring. Therefore, the constitutive relationship is postulated as follows:

$$\boldsymbol{F}_t(s) = \mathbf{C}'\boldsymbol{\delta}_t(s), \tag{2.7}$$

where the carcass stiffness matrix $\mathbf{C}'$ is given by

$$\mathbf{C}' = \begin{bmatrix} C'_{xx} & C'_{xy} \\ C'_{yx} & C'_{yy} \end{bmatrix}. \tag{2.8}$$

The matrix $\mathbf{C}'$ may be also assumed to be positive definite. Inverting Eq. (2.7) yields

$$\boldsymbol{\delta}_t(s) = \mathbf{S}'\boldsymbol{F}_t(s), \tag{2.9}$$

where $\mathbf{S}' \triangleq \mathbf{C}'^{-1}$ is the carcass compliance matrix.

2.3 Tyre-Road Kinematic Equations

In the brush theory, the governing equations of the tyre-road kinematics are two coupled partial differential equations (PDEs), which relate the micro-sliding velocity of the bristles to their rate of deflection. These equations may be derived starting from the theory of rolling contact mechanics developed by Kalker [18], which fairly applies to three-dimensional rigid bodies. Additional effects originating from tyre elasticity, such as carcass distortions related to conicity-induced effects, may be incorporated in the camber reduction factor ε_γ [1, 19]. The complete derivation is not reported here and may be found in [1, 20].

2.3.1 Exact Brush Theory for Large Camber Angles

Neglecting the deformation of the tyre carcass, the governing PDEs of the tyre-road kinematics may be cast in general form as

$$\bar{\boldsymbol{v}}_{\mathrm{S}}(\boldsymbol{x}, s) = -\boldsymbol{\sigma}(s) - \mathbf{A}_\varphi(s)\big(\boldsymbol{x} + \chi_\psi(s)\boldsymbol{u}_t(\boldsymbol{x}, s)\big) + \frac{\partial \boldsymbol{u}_t(\boldsymbol{x}, s)}{\partial s} + \big(\bar{\boldsymbol{v}}_t(\boldsymbol{x}, s) \cdot \nabla_t\big)\boldsymbol{u}_t(\boldsymbol{x}, s),$$

$$(\boldsymbol{x}, s) \in \mathring{\mathscr{P}} \times \mathbb{R}_{>0}, \tag{2.10}$$

where the tangential gradient operator reads

$$\nabla_t \triangleq \left[\partial/\partial x \ \partial/\partial y\right]^{\mathrm{T}}, \tag{2.11}$$

and the complete expression for the nondimensional velocity $\bar{\boldsymbol{v}}_t(\boldsymbol{x}, s)$ is as follows:

$$\bar{\boldsymbol{v}}_t(\boldsymbol{x}, s) \triangleq \frac{\mathrm{d}\boldsymbol{x}(s)}{\mathrm{d}s} = -\begin{bmatrix} 1 \\ 0 \end{bmatrix} + \mathbf{A}_{\varphi_\gamma}(s)\boldsymbol{x}. \tag{2.12}$$

In Eqs. (2.10) and (2.12), the matrices $\mathbf{A}_{\varphi}(s)$ and $\mathbf{A}_{\varphi_\gamma}(s)$ are called *spin* and *camber spin tensor*, respectively, and are given by [21]

$$\mathbf{A}_{\varphi}(s) \triangleq \begin{bmatrix} 0 & -\varphi(s) \\ \varphi(s) & 0 \end{bmatrix}, \tag{2.13a}$$

$$\mathbf{A}_{\varphi_\gamma}(s) \triangleq \begin{bmatrix} 0 & \varphi_\gamma(s) \\ -\varphi_\gamma(s) & 0 \end{bmatrix}. \tag{2.13b}$$

It is worth remarking that the nondimensional velocity $\bar{\boldsymbol{v}}_t(\boldsymbol{x}, s)$ in Eq. (2.12) represents the derivative of the coordinate $\boldsymbol{x}$ with respect to the travelled distance s. Therefore, integrating Eq. (2.12) provides the trajectories of the roots of the bristles travelling inside the contact patch. This aspect will be clarified in Chap. 5. According to Eq. (2.12), when the camber spin φ_γ is constant, that is for constant camber angles, these trajectories are circles centred in the cambering centre C_γ [21, 22]. Equations (2.10) and (2.12) are referred to in this book as *exact brush theory*, since, for a tyre with rigid carcass, they describe the exact kinematics of the bristles.

Equation (2.10) may be solved either for the deflection $\boldsymbol{u}_t(\boldsymbol{x}, s)$ or for the nondimensional micro-sliding velocity $\bar{\boldsymbol{v}}_{\mathrm{s}}(\boldsymbol{x}, s)$. In the adhesion zone $\mathscr{P}^{(\mathrm{a})}$, $\bar{\boldsymbol{v}}_{\mathrm{s}}(\boldsymbol{x}, s) = \mathbf{0}$ by definition, and Eq. (2.10) may be restated as

$$\frac{\partial \boldsymbol{u}_t^{(\mathrm{a})}(\boldsymbol{x}, s)}{\partial s} + \big(\bar{\boldsymbol{v}}_t(\boldsymbol{x}, s) \cdot \nabla_t\big)\boldsymbol{u}_t^{(\mathrm{a})}(\boldsymbol{x}, s) = \boldsymbol{\sigma}(s) + \mathbf{A}_{\varphi}(s)\Big(\boldsymbol{x} + \chi_{\psi}(s)\boldsymbol{u}_t^{(\mathrm{a})}(\boldsymbol{x}, s)\Big),$$

$$(\boldsymbol{x}, s) \in \mathring{\mathscr{P}}^{(\mathrm{a})} \times \mathbb{R}_{>0}. \tag{2.14}$$

The solution to the above PDE (2.14) is valid until the condition on the right-hand side of Eq. (2.2a) is satisfied. On the other hand, the bristle deflection $\boldsymbol{u}_t^{(\mathrm{s})}(\boldsymbol{x}, s)$ in the sliding zone $\mathscr{P}^{(\mathrm{s})}$ may be calculated from Eqs. (2.5) and (2.6) by substituting the expression (2.10) for the nondimensional micro-sliding velocity in Eq. (2.2b).

2.3.2 Classic Brush Theory

In the *classic brush theory*, the nondimensional velocity field $\bar{\boldsymbol{v}}_t(\boldsymbol{x}, s)$ inside the contact patch is approximated as $\bar{\boldsymbol{v}}_t(\boldsymbol{x}, s) \approx \bar{\boldsymbol{v}}_t = -\hat{\boldsymbol{e}}_x$. This simplification is acceptable for small camber angles, that is $\varphi_\gamma \approx 0$. Basically, in the classic brush theory, the trajectories of the bristles travelling inside the contact patch are straightened in the longitudinal direction. Accordingly, Eq. (2.10) reduces to

$$\bar{\boldsymbol{v}}_{\mathrm{s}}(\boldsymbol{x}, s) = -\boldsymbol{\sigma}(s) - \mathbf{A}_\varphi(s)\boldsymbol{x} + \frac{\partial \boldsymbol{u}_t(\boldsymbol{x}, s)}{\partial s} - \frac{\partial \boldsymbol{u}_t(\boldsymbol{x}, s)}{\partial x}, \quad (\boldsymbol{x}, s) \in \mathring{\mathscr{P}} \times \mathbb{R}_{>0}. \tag{2.15}$$

The corresponding formulation of (2.15) in the adhesion zone $\mathscr{P}^{(\mathrm{a})}$ is

$$\frac{\partial \boldsymbol{u}_t^{(\mathrm{a})}(\boldsymbol{x}, s)}{\partial s} - \frac{\partial \boldsymbol{u}_t^{(\mathrm{a})}(\boldsymbol{x}, s)}{\partial x} = \boldsymbol{\sigma}(s) + \mathbf{A}_\varphi(s)\boldsymbol{x}, \quad (\boldsymbol{x}, s) \in \mathring{\mathscr{P}}^{(\mathrm{a})} \times \mathbb{R}_{>0}. \tag{2.16}$$

2.3.3 Brush Models with Compliant Carcass

When the compliance of the tyre carcass is taken into account, the actual contact point C and the origin O of the contact patch may be displaced by an additional quantity $\boldsymbol{\delta}_t(s)$. In this case, the sliding velocity of the contact patch becomes $\boldsymbol{V}'_{\mathrm{s}}(s)$. Neglecting secondary effects due to the tyre carcass deflection, Eqs. (2.15) and (2.16) may be modified accordingly by replacing the slip variable $\boldsymbol{\sigma}(s)$ with the transient slip $\boldsymbol{\sigma}'(s)$.

2.4 Boundary and Initial Conditions

Equations (2.10) and (2.15) are two *linear transport equations* (first-order PDEs), defined on the interior $\mathring{\mathscr{P}}$ of the contact patch, and for positive values of the travelled distance s. To solve them, it is necessary to prescribe proper boundary conditions (BC) and initial conditions (IC). The BC should be prescribed on the boundaries of the adhesion and sliding zone $\partial\mathscr{P}^{(\mathrm{a})}$ and $\partial\mathscr{P}^{(\mathrm{s})}$, respectively. The IC should be instead prescribed on the initial configuration of their interiors $\mathring{\mathscr{P}}_0^{(\mathrm{a})} \triangleq \mathring{\mathscr{P}}^{(\mathrm{a})}(0)$ and $\mathring{\mathscr{P}}_0^{(\mathrm{s})} \triangleq \mathring{\mathscr{P}}^{(\mathrm{s})}(0)$.

2.4.1 Boundary and Initial Conditions for the Adhesion Zone

To formalise the BCs correctly, it is firstly necessary to introduce the notions of *leading edge* $\mathscr{L}$, *neutral edge* $\mathscr{N}$ and *trailing edge* $\mathscr{T}$ as follows [21]:

$$\mathscr{L} \triangleq \left\{ \boldsymbol{x} \in \partial\mathscr{P} \;\middle|\; \left[\bar{\boldsymbol{v}}_t(\boldsymbol{x}, s) - \bar{\boldsymbol{v}}_{\partial\mathscr{P}}(\boldsymbol{x}, s)\right] \cdot \hat{\boldsymbol{\nu}}_{\partial\mathscr{P}}(\boldsymbol{x}, s) < 0 \right\}, \tag{2.17a}$$

$$\mathscr{N} \triangleq \left\{ \boldsymbol{x} \in \partial\mathscr{P} \;\middle|\; \left[\bar{\boldsymbol{v}}_t(\boldsymbol{x}, s) - \bar{\boldsymbol{v}}_{\partial\mathscr{P}}(\boldsymbol{x}, s)\right] \cdot \hat{\boldsymbol{\nu}}_{\partial\mathscr{P}}(\boldsymbol{x}, s) = 0 \right\}, \tag{2.17b}$$

$$\mathscr{T} \triangleq \left\{ \boldsymbol{x} \in \partial\mathscr{P} \;\middle|\; \left[\bar{\boldsymbol{v}}_t(\boldsymbol{x}, s) - \bar{\boldsymbol{v}}_{\partial\mathscr{P}}(\boldsymbol{x}, s)\right] \cdot \hat{\boldsymbol{\nu}}_{\partial\mathscr{P}}(\boldsymbol{x}, s) > 0 \right\}, \tag{2.17c}$$

where $\hat{\boldsymbol{\nu}}_{\partial\mathscr{P}}(\boldsymbol{x}, s)$ is the outward-pointing unit normal to $\partial\mathscr{P}$ and $\bar{\boldsymbol{v}}_{\partial\mathscr{P}}(\boldsymbol{x}, s)$ is the velocity of the boundary of the contact patch $\partial\mathscr{P}$. The scalar product $[\bar{\boldsymbol{v}}_t(\boldsymbol{x}, s) - \bar{\boldsymbol{v}}_{\partial\mathscr{P}}(\boldsymbol{x}, s)] \cdot \hat{\boldsymbol{\nu}}_{\partial\mathscr{P}}(\boldsymbol{x}, s)$ represents the flow of the bristles through the boundary $\partial\mathscr{P}$ of the contact patch. It is worth emphasising that Eqs. (2.17a), (2.17b) and (2.17c) presume the existence of the unit normal. If $\partial\mathscr{P}$ is C^1, the unit normal may always be defined.[2] A direct consequence of the pure elastic constitutive relationship (2.5) is that adhesion conditions may start at the leading edge, where the bristles enter the contact patch. In the brush theory, it is assumed that the shear stress vanishes on the leading edge. Owing to Eq. (2.17a), this BC may be restated in mathematical terms as

$$\text{BC:} \quad \boldsymbol{q}_t(\boldsymbol{x}, s) = \mathbf{K}_t \boldsymbol{u}_t(\boldsymbol{x}, s) = \mathbf{0} \iff \boldsymbol{u}_t(\boldsymbol{x}, s) = \mathbf{0}, \quad (\boldsymbol{x}, s) \in \mathscr{L} \times \mathbb{R}_{>0}, \tag{2.18}$$

since $\mathbf{K}_t$ is nonsingular.

Additionally, an adhesion zone may also originate from a previous sliding zone. In this case, the BC must be prescribed on an *adhesion edge* $\mathscr{A}$ and may be formulated as follows:

$$\text{BC:} \qquad \boldsymbol{u}_t^{(\mathrm{a})}(\boldsymbol{x}, s) = \boldsymbol{u}_t^{(\mathrm{s})}(\boldsymbol{x}, s), \qquad (\boldsymbol{x}, s) \in \mathscr{A} \times \mathbb{R}_{>0}. \tag{2.19}$$

The above BC (2.19) prescribes the continuity of the solution on the adhesion edge $\mathscr{A}$. Finally, a general IC may formulated mathematically as[3]

$$\text{IC:} \qquad \boldsymbol{u}_t^{(\mathrm{a})}(\boldsymbol{x}, 0) = \boldsymbol{u}_{t0}^{(\mathrm{a})}(\boldsymbol{x}), \qquad \boldsymbol{x} \in \mathring{\mathscr{P}}_0^{(\mathrm{a})}, \tag{2.20}$$

---

[2] When $\partial\mathscr{P}$ is only C^0, the solution may be not well defined on the trajectories that originate at the corners. However, this does not introduce any complication in the computation of the tyre forces and moment, since these lines have the Lebesgue measure zero. See also [21].

[3] As also discussed in [21], the assumption $\boldsymbol{u}_{t0}^{(\mathrm{a})}(\boldsymbol{x}) \in C^1(\mathring{\mathscr{P}}_0^{(\mathrm{a})}; \mathbb{R}^2)$ is not obvious, and often only weak solutions may be found to Eqs. (2.10) and (2.15).

for some $\boldsymbol{u}_{t0}^{(a)}(\boldsymbol{x}) \in C^1(\mathring{\mathscr{P}}_0^{(a)}; \mathbb{R}^2)$ with $\boldsymbol{u}_{t0}^{(a)}(\boldsymbol{x}) = \boldsymbol{0}$ on $\mathscr{L}_0 \triangleq \mathscr{L}(0)$ and continuous with a previous sliding solution on $\mathscr{S}_0 \triangleq \mathscr{S}(0)$.

2.4.2 *Boundary Condition for the Sliding Zone*

The last BC that needs to be introduced concerns the transition from adhesion to sliding. This happens on a *sliding edge* $\mathscr{S}$, which separates two regions $\mathscr{P}^{(a)}$ and $\mathscr{P}^{(s)}$, marking the transition from Eq. (2.2a) to (2.2b). To formulate the corresponding BC, it may be beneficial to define

$$\gamma_{\mathscr{S}}(\boldsymbol{x}, s) \triangleq \left\| \mathbf{K}_t \boldsymbol{u}_t^{(a)}(\boldsymbol{x}, s) \right\| - \mu q_z(\boldsymbol{x}, s), \tag{2.21}$$

where the adhesion solution $\boldsymbol{u}_t^{(a)}(\boldsymbol{x}, s)$ is described by a known function that does not coincide locally with the friction limit $\mu q_z(\boldsymbol{x}, s)$. According to Eq. (2.21), a sliding edge may be represented implicitly as

$$\mathscr{S} \triangleq \left\{ \boldsymbol{x} \in \mathscr{P} \mid \gamma_{\mathscr{S}}(\boldsymbol{x}, s) = 0 \right\}. \tag{2.22}$$

The BC for the solution $\boldsymbol{u}_t^{(s)}(\boldsymbol{x}, s)$ in the sliding zone $\mathscr{P}^{(s)}$ is, thus, formulated as follows:

$$\text{BC:} \quad \boldsymbol{u}_t^{(s)}(\boldsymbol{x}, s) = \mathbf{K}_t^{-1} \mu q_z(\boldsymbol{x}, s) \hat{\boldsymbol{s}}_t(\boldsymbol{x}, s) = \boldsymbol{u}_t^{(a)}(\boldsymbol{x}, s), \quad (\boldsymbol{x}, s) \in \mathscr{S} \times \mathbb{R}_{>0}. \tag{2.23}$$

Similarly to (2.19), the above BC (2.23) enforces the continuity of the solution in the transition from adhesion to sliding. Eventually, if needed, an IC for the sliding zone might be cast as done in Eq. (2.20).

2.5 Equilibrium Equations

The equilibrium equations establish the relationships between the local shear stresses arising in the contact patch and the forces and moment acting on the tyre, usually calculated with respect to the contact patch centre. If the carcass is assumed to be rigid, this practically coincides with the virtual contact point. In terms of theoretical slip variables $(\boldsymbol{\sigma}', \varphi)$, the tyre characteristics may be found by integration[4] over the contact patch as follows:

---

[4] When integrating over $\mathscr{P}$, it may be written, with some abuse of notation, $\mathrm{d}\boldsymbol{x} = \mathrm{d}x\,\mathrm{d}y$, since z is fixed.

$$\boldsymbol{F}_t(\boldsymbol{\sigma}, \boldsymbol{\varphi}, s) = \iint_{\mathscr{P}(s)} \boldsymbol{q}_t\left(\boldsymbol{x}, s; \boldsymbol{\sigma}, \boldsymbol{\varphi}\right) \mathrm{d}\boldsymbol{x}, \tag{2.24a}$$

$$M_z(\boldsymbol{\sigma}, \boldsymbol{\varphi}, s) = \iint_{\mathscr{P}(s)} \big(x + u_x(\boldsymbol{x}, s)\big) q_y(\boldsymbol{x}, s; \boldsymbol{\sigma}, \boldsymbol{\varphi})\,\mathrm{d}\boldsymbol{x}$$
$$- \iint_{\mathscr{P}(s)} \big(y + u_y(\boldsymbol{x}, s)\big) q_x(\boldsymbol{x}, s; \boldsymbol{\sigma}, \boldsymbol{\varphi})\,\mathrm{d}\boldsymbol{x}. \tag{2.24b}$$

In Eqs. (2.24a), (2.24b), the bristle deflection $\boldsymbol{u}_t(\boldsymbol{x}, s)$ represents the solution to the PDEs (2.10) or (2.15), and the shear stress $\boldsymbol{q}_t(\boldsymbol{x}, s)$ is calculated according to Eq. (2.5). Often, in (2.24b), the equilibrium is calculated on the undeformed configuration, that is neglecting the bristle displacements in the levers by setting $x + u_x(\boldsymbol{x}, s) \simeq x$ and $y + u_y(\boldsymbol{x}, s) \simeq y$. It is necessary to clarify that the moment $M_z \equiv M_z^{(O)}$ in Eq. (2.24b) is calculated with respect to the contact-fixed point O. For small camber angles and a rigid tyre carcass, $O \approx C \approx B$. More generally, when $O \not\equiv B$, the self-aligning moment computed with respect to the virtual contact B point reads

$$M_z^{(B)}(\boldsymbol{\sigma}, \boldsymbol{\varphi}, s) = M_z(\boldsymbol{\sigma}, \boldsymbol{\varphi}, s) - \big[c_{\mathrm{r}}(\gamma) + \delta_y(s)\big] F_x(\boldsymbol{\sigma}, \boldsymbol{\varphi}, s) + \delta_x(s) F_y(\boldsymbol{\sigma}, \boldsymbol{\varphi}, s). \tag{2.25}$$

In the following chapters, the superscripts $(\cdot)^{(B)}$ and $(\cdot)^{(O)}$ will be omitted for brevity, and M_z will be used to denote the self-aligning moment computed with respect to O.

It is also interesting to observe that, in transient conditions, above Eqs. (2.24a), (2.24b) yield an implicit description of the tyre dynamics as in Eq. (1.16), since $\boldsymbol{\sigma}'$ is, in turn, a function of $\boldsymbol{F}_t$ according to Eqs. (1.5) and (2.9), and so are the shear stresses. On the other hand, in steady-state conditions, the shear stresses do not depend upon the travelled distance s, and moreover, $\boldsymbol{\sigma}' \equiv \boldsymbol{\sigma}$. Therefore, Eqs. (2.24a), (2.24b) yield an explicit representation of the tyre forces and moment in the form (1.22). The equivalent descriptions in terms of practical or geometrical slip variables $(\boldsymbol{\kappa}, \boldsymbol{\varphi})$ or $(\kappa_x, \alpha, \gamma, \varphi_\psi)$ may always be obtained from Eqs. (2.24a), (2.24b) with the aid of the opportune transformations introduced in Sect. 1.2.

References

1. Pacejka HB (2012) Tire and vehicle dynamics, 3rd edn. Elsevier/BH, Amsterdam
2. Duvaut G, Lions JL (1976) Inequalities in mechanics and physics. Springer, Berlin Heidelberg. https://doi.org/10.1007/978-3-642-66165-5
3. Kalker JJ (1997) Variational principles of contact elastostatics. J Inst Math Its Appl 20:199–219
4. Guiggiani M (2018) The science of vehicle dynamics, 2nd edn. Springer International, Cham (Switzerland)
5. Canudas de Wit C, Olsson H, Astrom KJ, Lischinsky P (1995) A new model for control of systems with friction. IEEE Trans Autom Control 40(3):419–425. https://doi.org/10.1109/9.376053

6. Sharifzadeh M, Timpone F, Farnam A, Senatore A, Akbari A (2017) Tyre-road adherence conditions estimation for intelligent vehicle safety applications. In: Advances in Italian mechanism science. Adv Ital Mech Science Mech Mach Sci, vol 47. Springer, Cham. https://doi.org/10.1007/978-3-319-48375-7_42
7. Sharifzadeh M, Senatore A, Farnam A, Akbari A, Timpone F (2019) A real-time approach to robust identification of tyre–road friction characteristics on mixed-¯ roads. Vehicle Syst Dyn 57(9):1338–1362. https://doi.org/10.1080/00423114.2018.1504974
8. Canudas-de-Wit C, Tsiotras P, Velenis E, Basset M, Gissinger G (2003) Dynamic friction models for road/tire longitudinal interaction. Veh Syst Dyn 39(3)189–226. https://doi.org/10.1076/esd.39.3.189.14152 https://doi.org/10.1076/vesd.39.3.189.14152
9. Deur J, Ivanovic V, Troulis M, Miano C, Hrovat D, Asgari J (2005) Extensions of the LuGre tyre friction model related to variable slip speed aong the contact patch length. Veh Syst Dyn 43(supp):508–524. https://doi.org/10.1080/00423110500229808
10. Deur J, Asgari J, Hrovat D (2004) A 3D brush-type dynamic tire friction model. Veh Syst Dyn 42(3):133–173. https://doi.org/10.1080/00423110412331282887
11. Velenis E, Tsiotras P, Canudas-de-Wit C, Sorine M (2005) Dynamic tyre friction models for combined longitudinal and lateral vehicle motion. Veh Syst Dyn 43(1):3–29. https://doi.org/10.1080/00423110412331290464
12. Liang W, Medanic J, Ruhl R (2008) Analytical dynamic tire model. Veh Syst Dyn 46(3):197–227. https://doi.org/10.1080/00423110701267466
13. Persson BNJ (2001) Theory of rubber friction and contact mechanics. J Chem Phys 115(8):3840–3861. https://doi.org/10.1063/1.1388626
14. Persson BNJ, Albohr O, Tartaglino U, Volokitin AI, Tosatti E (2004) On the nature of surface roughness with application to contact mechanics, sealing, rubber friction and adhesion. J Phys: Condens Matter 17(1):R1–R62
15. Persson BNJ (2006) Contact mechanics for randomly rough surfaces. Surf Sci Rep 61(4):201–227. https://doi.org/10.1016/j.surfrep.2006.04.001
16. O'Neill A, Gruber P, Watts JF, Prins J (2019) Predicting tyre behaviour on different road surfaces. In: Klomp M, Bruzelius F, Nielsen J, Hillemyr A (eds) Advances in dynamics of vehicles on roads and tracks. IAVSD. Lecture notes in mechanical engineering. Springer, Cham. https://doi.org/10.1007/978-3-030-38077-9_215
17. O'Neill A, Prins F, Watts JF, Gruber P (2021) Enhancing brush tyre model accuracy through friction measurements. Veh Sys Dyn. https://doi.org/10.1080/00423114.2021.1893766
18. Kalker JJ (1990) Three-dimensional elastic bodies in rolling contact. Springer, Dordrecht. https://doi.org/10.1007/978-94-015-7889-9
19. Pacejka HB (2005) Spin: camber and turning. Veh Syst Dyn 43(1):3–17. https://doi.org/10.1080/00423110500140013
20. Limebeer DJN, Massaro M (2018) Dynamics and optimal control of road vehicle. Oxford University Press
21. Romano L, Timpone F, Bruzelius F, Jacobson B. Analytical results in transient brush tyre models: theory for large camber angles and classic solutions with limited friction. Meccanica. https://doi.org/10.1007/s11012-021-01422-3
22. Romano L, Bruzelius F, Jacobson B. Brush tyre models for large camber angles and steering speeds. Vehicle Syst Dyn. https://doi.org/10.1080/00423114.2020.1854320

Chapter 3
Steady-State Brush Theory

Abstract In steady-state conditions, explicit expressions for tyre characteristics may be derived using the theoretical framework provided by the brush theory. This chapter is, thus, dedicated to addressing the stationary problem from both the local and global perspectives. The fundamental concepts of critical slip and spin are introduced with respect to an isotropic tyre, and the deformation of the bristles inside the contact patch is investigated for different operating conditions of the tyre. Analytical functions describing the tyre forces and moment acting inside the contact patch are obtained for the particular case of a rectangular contact patch. The analysis is qualitative in nature.

In practical applications, the tyre characteristics are often described quantitatively using *ad-hoc* developed models. In particular, the so-called Pacejka's Magic Formula (MF) [1, 5] represents nowadays the standard approach when it comes to simulation and optimal control of road vehicles. The earliest version of this empirical model was an outcome of a collaboration between Volvo and Delft University of Technology, and, at the time, combined slip conditions were still handled from a physical viewpoint. Later on, a better agreement with experimental data was achieved by employing pure empirical relationships to describe the tyre forces [6]. Over the years, the MF tyre model has been refined progressively primarily by Pacejka and Besselink [7, 8], who extended the original formulation to deal with the presence of large camber angles, variations in the inflation pressure and transient dynamics. At present, the MF model collects around eighty macro and micro-parameters and is able to fit with surgical precision almost any input. It has been recently shown in [9–13] that these Pacejka's MF coefficients may eventually be interpreted as functional parameters, which allow accounting for thermal and wear-related phenomena. Albeit being empirical in nature, the MF takes inspiration from the brush theory to properly model certain effects. Indeed, as a qualitative description, the brush models provide an intuitive understanding of the fundamental phenomena governing the tyre-road interaction.

In this context, it is worth remarking that even the most advanced tyre models requiring computer simulations, like, for example, FTire® [14–16] and CDTire [17],

L. Romano, *Advanced Brush Tyre Modelling*,
SpringerBriefs in Applied Sciences and Technology,
https://doi.org/10.1007/978-3-030-98435-9_3

take advantage of the brush theory to provide a more realistic description of the local contact phenomena occurring between the tyre tread and the road. In particular, FTire® represents the most sophisticated model based on a nonlinear beam formulation and incorporates additional features such as belt compliance and distributed spring-like elements to properly capture the tread deformation. CDTire is based instead on a 3D shell description of both the sidewall and the belt, but also includes a dedicated brush-type contact model. Currently, they are both extensively used in the context of advanced driving simulations or for comfort applications. Indeed, despite their complexity, FTire® and CDTire are capable of running in real-time.

From the brief *excursus* above, it should be clear that, due to its eminent role and its relative simplicity, the brush theory should be generally regarded as the main theoretical foundation to the development of any other tyre model. Thus, this chapter is dedicated to the classic stationary theory of the brush models. The overall objective is to exploit the results from Chap. 2 to derive closed-form expressions for the steady-state tyre characteristics as a function of the theoretical slips as in Eqs. (1.22). The equivalent formulations in terms of practical and geometrical slips may be then deduced with the aid of the transformations introduced in Sect. 1.2.

The results advocated in the following are also propaedeutic to the subsequent investigations conducted in Chaps. 4 and 6, where the transient phenomena concerning the tyre-road interaction are explained within the theoretical framework of the classic brush theory.

3.1 Assumptions and Hypotheses

The exposition is grounded on the two following technical Assumptions 3.1.1 and 3.1.2 [18].

Assumption 3.1.1 The contact patch $\mathscr{P}$ is a compact, D-convex[1] set along the direction $\hat{\boldsymbol{e}}_x$.

The direction $\hat{\boldsymbol{e}}_x$ is parallel to the nondimensional velocity field inside the contact patch, which, in the classic brush theory, is given by $\bar{\boldsymbol{v}}_t = -\hat{\boldsymbol{e}}_x$. Two typical geometries satisfying Assumption 3.1.1 are the rectangular contact patch

$$\mathscr{P} \triangleq \left\{ \boldsymbol{x} \in \Pi \;\middle|\; -a \le x \le a,\ -b \le y \le b \right\}, \tag{3.1}$$

and the elliptical contact patch

[1] A D-convex set is a set convex along a specified direction. Analogously, a D-convex function is a function which is convex along a given direction. The reader may refer to [19] for additional details.

$$\mathscr{P} \triangleq \left\{ \boldsymbol{x} \in \Pi \;\middle|\; \frac{x^2}{a^2} + \frac{y^2}{b^2} \leq 1 \right\}. \tag{3.2}$$

Furthermore, Assumption 3.1.1 ensures the existence of a continuous leading edge[2].

Example 3.1.1 For a rectangular contact patch, the leading, neutral and trailing edges may be parametrised, respectively, by

$$x = x_{\mathscr{L}}(y) \triangleq a, \qquad y \in (-b, b), \tag{3.3a}$$

$$y = y_{\mathscr{N}}(x) \triangleq \pm b, \qquad x \in (-a, a), \tag{3.3b}$$

$$x = x_{\mathscr{T}}(y) \triangleq -a, \qquad y \in (-b, b). \tag{3.3c}$$

It is worth noticing that for points located at $(\pm a, \pm b)$ the unit normal to $\partial\mathscr{P}$ is not defined. On the other hand, for an elliptical contact patch,

$$x = x_{\mathscr{L}}(y) \triangleq a\sqrt{1 - \frac{y^2}{b^2}}, \qquad y \in (-b, b), \tag{3.4a}$$

$$x = x_{\mathscr{T}}(y) \triangleq -a\sqrt{1 - \frac{y^2}{b^2}}, \qquad y \in (-b, b), \tag{3.4b}$$

and there are only two neutral points located at $\boldsymbol{x}_{\mathscr{N}} = (0, \pm b)$.

To investigate the classic theory, it may also be useful to introduce a change of coordinates. If the leading and trailing are parametrised by $x = x_{\mathscr{L}}(y)$ and $x = x_{\mathscr{T}}(y)$, it is possible to define

$$\boldsymbol{\xi} = \begin{bmatrix} \xi \\ \eta \\ \zeta \end{bmatrix} \triangleq \begin{bmatrix} x_{\mathscr{L}}(y) - x \\ y \\ z \end{bmatrix} \tag{3.5a}$$

and replace the original coordinates using the transformation $\boldsymbol{x} \mapsto \boldsymbol{\xi}$. It should be noticed that the variable ξ is a local coordinate and represents the distance from the leading edge (parametrised explicitly by $\xi = \xi_{\mathscr{L}}(\eta) = 0$) for a fixed η. For this reason, it is often referred to as *distance from the entrance*. Apart from having a clear physical meaning, however, the transformation introduced in Eq. (3.5) allows to *straighten out the boundary*, as discussed in [20]: in these coordinates, the patch becomes a rectangle 'in the front', with a possibly non-straight line for the trailing edge. Using the coordinates $\boldsymbol{\xi}$, the trailing edge may be parametrised explicitly as $\xi = \xi_{\mathscr{T}}(\eta)$. The rectangular and elliptical contact patches may be described in the coordinate system $\boldsymbol{\xi}$ as shown below.

[2] In turn, the existence of a continuous leading edge ensures that the steady-state displacement of the bristle is at least $C^0(\mathscr{P})$.

Example 3.1.2 Replacing the original coordinates $\boldsymbol{x}$ with $\boldsymbol{\xi}$, both the rectangular and elliptical contact patches described mathematically by Eqs. (3.1) and (3.2) may be restated as

$$\mathscr{P} \triangleq \left\{\boldsymbol{\xi} \in \Pi \;\middle|\; 0 \le \xi \le \xi_{\mathscr{T}}(\eta),\ -b \le \eta \le b\right\}, \tag{3.6}$$

where $\xi_{\mathscr{T}}(\eta)$ is a parametrisation of the trailing edge $\xi = \xi_{\mathscr{T}}(\eta)$ in the coordinates $\boldsymbol{\xi}$. More specifically, for a rectangular contact patch,

$$\xi = \xi_{\mathscr{T}}(\eta) \triangleq 2a, \qquad \eta \in (-b, b), \tag{3.7}$$

whilst for an elliptical contact patch,

$$\xi = \xi_{\mathscr{T}}(\eta) \triangleq 2a\sqrt{1 - \frac{\eta^2}{b^2}}, \qquad \eta \in (-b, b). \tag{3.8}$$

The following Assumption 3.1.2 restricts the attention to a particular class of pressure distributions inside the contact patch $\mathscr{P}$ [18].

Assumption 3.1.2 For every $\boldsymbol{\xi} \in \mathscr{P}$, consider the restrictions of the contact patch and vertical pressure distribution obtained for η fixed, i.e. $\mathscr{P}^{(\eta)} \triangleq \mathscr{P} \restriction_{\eta}$ and $q_z^{(\eta)}(\xi) \triangleq q_z(\boldsymbol{\xi}) \restriction_{\eta}$. It is assumed that $q_z^{(\eta)} \in C^1(\mathscr{P}^{(\eta)}; \mathbb{R})$, with $q_z^{(\eta)}(\xi) = 0$ on $\partial\mathscr{P}^{(\eta)}$ and $q_z^{(\eta)}(\cdot)$ concave (i.e. $q_z(\cdot)$ D-concave in direction $\hat{\boldsymbol{e}}_x$).

Some examples of pressure distributions satisfying above Assumption 3.1.2 are given below.

Example 3.1.3 *(Rectangular and elliptical contact patches)* For a rectangular and elliptical contact patch, the pressure distribution is often assumed to be of the form

$$q_z(\boldsymbol{\xi}) = \frac{q_z^*}{a^2}\xi\big(\xi_{\mathscr{T}}(\eta) - \xi\big), \tag{3.9}$$

with $\xi_{\mathscr{T}}(\eta)$ reading, respectively, as in Eqs. (3.7) and (3.8). The restriction $q_z^{(\eta)}(\xi)$ of a pressure distribution as in Eq. (3.22) is clearly strictly convex and attains zero values for $\xi = \xi_{\mathscr{L}}(\eta) = 0$ and $\xi = \xi_{\mathscr{T}}(\eta)$, which correspond to the functions that parametrise the leading and trailing edges.

3.2 Vanishing Sliding

According to the classic brush theory [1], the deflection of the bristle in steady-state conditions ($\boldsymbol{\sigma}'(s) \equiv \boldsymbol{\sigma}(s)$) is governed by the following PDEs:

$$\bar{\boldsymbol{v}}_{\mathrm{s}}(\boldsymbol{\xi}) = -\boldsymbol{\sigma} - \mathbf{A}_{\varphi}\begin{bmatrix} x_{\mathscr{L}}(\eta) - \xi \\ \eta \end{bmatrix} + \frac{\partial \boldsymbol{u}_t(\boldsymbol{\xi})}{\partial \xi}, \qquad \boldsymbol{\xi} \in \mathring{\mathscr{P}}, \tag{3.10}$$

which may be derived from Eq. (2.15) using the transformation (3.5) defined above. As a first step, the problem may be analysed under the assumption of vanishing sliding, that is $\bar{\boldsymbol{v}}_s = \mathbf{0}$ all over $\mathscr{P}$ (or, equivalently, $\mathscr{P}^{(a)} \equiv \mathscr{P}$). Such condition holds approximately true either when the translational and rotational slips are sufficiently small, or the friction available inside the contact patch is virtually infinite [1]. The latter situation translates mathematically into $\mu \to \infty$. In vanishing sliding, Eq. (3.10) becomes

$$\frac{\partial \boldsymbol{u}_t(\boldsymbol{\xi})}{\partial \xi} = \boldsymbol{\sigma} + \mathbf{A}_\varphi \begin{bmatrix} x_{\mathscr{L}}(\eta) - \xi \\ \eta \end{bmatrix}, \qquad \boldsymbol{\xi} \in \mathring{\mathscr{P}}, \tag{3.11}$$

and comes equipped with the following BC:

$$\text{BC:} \qquad \boldsymbol{u}_t(0, \eta) = \mathbf{0}. \tag{3.12}$$

The global solution $\boldsymbol{u}_t(\boldsymbol{\xi}) \in C^\infty(\mathscr{P}; \mathbb{R}^2)$ is given by

$$\boldsymbol{u}_t(\boldsymbol{\xi}) = \boldsymbol{\sigma}\xi + \mathbf{A}_\varphi \xi \begin{bmatrix} x_{\mathscr{L}}(\eta) - \xi/2 \\ \eta \end{bmatrix}, \qquad \boldsymbol{\xi} \in \mathscr{P}. \tag{3.13}$$

Integrating over the contact patch according to Eqs. (2.24) yields the tyre characteristics. In the classic brush theory, the equilibrium is actually computed on the undeformed configuration. The calculation may be worked analytically for a tyre with rectangular contact patch and isotropic tread with diagonal stiffness matrix, that is $k_{xx} = k_{yy} = k$ and $k_{xy} = k_{yx} = 0$ in Eq. (2.6). In this case, the following relationships are obtained:

$$\boldsymbol{F}_t(\boldsymbol{\sigma}, \varphi) = C_\sigma \boldsymbol{\sigma} + C_\varphi \varphi \hat{\boldsymbol{e}}_y, \tag{3.14a}$$

$$M_z(\sigma_y, \varphi) = -C_{M\sigma_y}\sigma_y + C_{M\varphi}\varphi, \tag{3.14b}$$

with

$$C_\sigma = 4ka^2b, \qquad C_\varphi \equiv C_{M\sigma_y} = \frac{a}{3}C_\sigma, \qquad C_{M\varphi} = \frac{b^2}{3}C_\sigma. \tag{3.15}$$

In Eq. (3.15), the stiffnesses have been renamed $C_\sigma \triangleq C_{x\sigma_x} \equiv C_{y\sigma_y}$ and $C_\varphi \triangleq C_{y\varphi}$ for the sake of notation and without ambiguity. It may be easily verified that the definition above coincides with that of theoretical slip stiffnesses introduced in Subsect. 1.4.1.

For an elliptical contact patch, the expressions in Eqs. (3.14) are retained, but with

$$C_\sigma = \frac{8}{3}ka^2b, \qquad C_\varphi \equiv C_{M\sigma_y} = \frac{3\pi a}{32}C_\sigma. \tag{3.16}$$

From Eqs. (3.14), the self-aligning moment may be expressed as

$$M_z(\sigma_y, \varphi) = -t_{\mathrm{p}}(\sigma_y, \varphi) F_y(\sigma_y, \varphi), \tag{3.17}$$

where the quantity t_{p} is called *pneumatic trail*. The negative sign in Eq. (3.17) makes t_{p} positive for normal operating conditions. For example, for a rectangular contact patch, the pneumatic trail may be easily computed as

$$t_{\mathrm{p}}(\sigma_y, \varphi) = \frac{a\sigma_y - b^2\varphi}{3\sigma_y + a\varphi}. \tag{3.18}$$

Equation (3.17) is particularly interesting, since it states that zero lateral force also implies zero self-aligning moment. It is important to remark that the self-aligning moment given by Eq. (3.14) is computed with respect to the origin O of the patch-fixed reference frame, which in this case coincides with its centre. The corresponding moment acting on the virtual contact point may be calculated by means of Eq. (2.25), provided that a relationship for $c_{\mathrm{r}}(\gamma)$ is known.

It is deduced from Eqs. (3.14) that, in vanishing sliding conditions, a linear relationship exists between the tyre forces and moment and the slip and spin variables. In case of limited friction available inside the contact patch, the assumption of vanishing sliding is not valid anymore, and a linear approximation similar to that in Eqs. (3.14) holds only for sufficiently small slips. For higher values of $\boldsymbol{\sigma}$ and φ, large sliding areas may arise inside the contact patch and Eq. (2.2b) needs to be considered. For the case under consideration, it may be restated as follows:

$$\boldsymbol{u}_t^{(\mathrm{s})}(\boldsymbol{\xi}, s) = \frac{\mu}{k} q_z(\boldsymbol{\xi}) \hat{\boldsymbol{s}}_t(\boldsymbol{\xi}, s), \tag{3.19}$$

where the sliding direction $\hat{\boldsymbol{s}}_t(\boldsymbol{\xi}, s)$ is defined as in Eq. (2.3) and the nondimensional micro-sliding velocity reads as in Eq. (3.10). The analysis in presence of limited friction is conducted in the next sections for the cases of pure translational slips, pure spin and combined lateral and spin slips.

3.3 Pure Translational Slip

In pure translational slip conditions, that is $\boldsymbol{\sigma} \neq \boldsymbol{0}$ and $\varphi = 0$, the analysis may be worked out analytically by limiting the focus to an isotropic tyre with a diagonal stiffness matrix. To this end, it may be beneficial to introduce the notion of *local critical slip* as follows:

$$\sigma^*(\eta) \triangleq \frac{\mu}{k} \left\| \frac{\partial q_z(\boldsymbol{\xi})}{\partial \xi} \right\|_{\xi=0}. \tag{3.20}$$

It may be observed that $\sigma^*(\eta)$ is a function of the lateral coordinate and hence is a local variable inside the contact patch. In absence of spin and if the pressure distribution satisfies Assumption 3.1.2, the sections $\eta = \text{const}$ of the contact patch where $\sigma \triangleq \|\boldsymbol{\sigma}\| > \sigma^*(\eta)$ are in full sliding conditions. In this context, it should be observed that, when $q_z^{(\eta)}(\cdot)$ in Assumption 3.1.2 is also strictly concave[3] for some η, full sliding occurs even for $\sigma \geq \sigma^*(\eta)$ for those η. From Eq. (3.20), it is still possible to define the *global critical slip value* which causes total sliding in the entire contact patch, that is for all η,

$$\sigma^{\text{cr}} \triangleq \sup_{\eta} \sigma^*(\eta). \tag{3.21}$$

Example 3.3.1 (Critical slip values for rectangular and elliptical contact patch) For a rectangular contact patch, the pressure distribution $q_z(\boldsymbol{\xi})$ may be recast explicitly as

$$q_z(\xi) = \frac{q_z^*}{a^2}\xi(2a - \xi), \tag{3.22}$$

and thus, from (3.20), it follows that

$$\sigma^{\text{cr}} \equiv \sigma^* = 2\mu\frac{q_z^*}{ka} = \frac{3\mu F_z}{C_\sigma}, \tag{3.23}$$

since σ^* is independent of the lateral coordinate η. In Eq. (3.23), the slip stiffness C_σ is defined as in (3.15). Conversely, for an elliptical contact patch,

$$\sigma^*(\eta) = 2\mu\frac{q_z^*}{ka}\sqrt{1 - \frac{\eta^2}{b^2}}, \tag{3.24a}$$

$$\sigma^{\text{cr}} \triangleq \sup_{\eta} \sigma^*(\eta) = 2\mu\frac{q_z^*}{ka} = \frac{32\mu F_z}{3\pi C_\sigma}, \tag{3.24b}$$

with C_σ reading as in (3.16). In particular, it may be observed that the maximum value of $\sigma^*(\eta)$ is attained for $\eta = 0$, and hence, the middle plane of the tyre is the last section to undergo full sliding.

The global solution $\boldsymbol{u}_t(\boldsymbol{\xi}) \in C^0(\mathscr{P}; \mathbb{R}^2)$ for the bristle displacement may be constructed as

$$\boldsymbol{u}_t(\boldsymbol{\xi}) = \begin{cases} \boldsymbol{u}_t^{(\text{a})}(\boldsymbol{\xi}), & \boldsymbol{\xi} \in \mathscr{P}^{(\text{a})}, \\ \boldsymbol{u}_t^{(\text{s})}(\boldsymbol{\xi}), & \boldsymbol{\xi} \in \mathscr{P}^{(\text{s})}, \end{cases} \tag{3.25}$$

where the adhesion solution $\boldsymbol{u}_t^{(\text{a})}(\xi)$ is obtained from Eq. (3.13) as

[3] From a global perspective, full and partial sliding conditions have not been defined rigorously in the literature. An intuitive way of stating mathematically full sliding conditions would be to consider the Lebesgue measures $\lambda^*(\cdot)$ of the contact patch $\mathscr{P}$ and its sliding region $\mathscr{P}^{(\text{s})}$. Then full sliding would occur if $\lambda^*(\mathscr{P}^{(\text{s})}) \equiv \lambda^*(\mathscr{P})$, and partial sliding if $\lambda^*(\mathscr{P}^{(\text{s})}) < \lambda^*(\mathscr{P})$. With the same rationale, vanishing sliding conditions would correspond to $\lambda^*(\mathscr{P}^{(\text{a})}) \equiv \lambda^*(\mathscr{P})$.

$$\boldsymbol{u}_t^{(\mathrm{a})}(\xi) = \boldsymbol{\sigma}\xi, \qquad \boldsymbol{\xi} \in \mathscr{P}^{(\mathrm{a})}, \tag{3.26}$$

whilst the sliding solution may be sought as follows:

$$\boldsymbol{u}_t^{(\mathrm{s})}(\boldsymbol{\xi}) = \frac{\mu}{k} q_z(\boldsymbol{\xi}) \frac{\boldsymbol{\sigma}}{\sigma} \qquad \text{if} \qquad k\sigma > \mu \frac{\partial q_z(\boldsymbol{\xi})}{\partial \xi}, \tag{3.27}$$

with $\sigma = \|\boldsymbol{\sigma}\|$. In Eq. (3.25), the adhesion and sliding zones $\mathscr{P}^{(\mathrm{a})}$ and $\mathscr{P}^{(\mathrm{s})}$, respectively, are separated by a sliding edge $\mathscr{S}$, which may sometimes be defined as $\mathscr{S} = \{\boldsymbol{\xi} \in \mathscr{P} \mid \xi - \xi_{\mathscr{S}}(\eta) = 0\}$, where $\xi_{\mathscr{S}}(\eta)$ is an explicit parametrisation of the sliding edge. From the definition of a sliding edge, $\xi_{\mathscr{S}}(\eta)$ may be found by equating the expression for the total adhesion shear stress $q_t^{(\mathrm{a})}(\xi) \triangleq \left\| \boldsymbol{q}_t^{(\mathrm{a})}(\xi) \right\| = k\sigma\xi$ to that of the friction bound $\mu q_z(\boldsymbol{\xi})$ and solving for ξ. For a rectangular contact patch with parabolic pressure distribution, the explicit parametrisation for the sliding edge may be calculated as

$$\xi_{\mathscr{S}}(\eta) = 2a\left(1 - \frac{\sigma}{\sigma^{\mathrm{cr}}}\right) \triangleq \lambda_{\mathrm{ad}}, \qquad \eta \in [-b, b] \ \text{ and } \ \sigma \le \sigma^{\mathrm{cr}}, \tag{3.28}$$

where the quantity λ_{ad} is often referred to as *adhesion length*. For an elliptical contact patch with parabolic pressure distribution, it is instead given by

$$\xi_{\mathscr{S}}(\eta) = 2a\left(\sqrt{1 - \frac{\eta^2}{b^2}} - \frac{\sigma}{\sigma^{\mathrm{cr}}}\right), \qquad \eta \in \left[-\eta^*, \eta^*\right] \ \text{ and } \ \sigma \le \sigma^*(\eta), \tag{3.29}$$

with

$$\eta^* \triangleq b\sqrt{1 - \left(\frac{\sigma}{\sigma^{\mathrm{cr}}}\right)^2}. \tag{3.30}$$

Therefore, in both cases, the adhesion and sliding zone may be restated as $\mathscr{P}^{(\mathrm{a})} = \{\boldsymbol{\xi} \in \mathscr{P} \mid \xi - \xi_{\mathscr{S}}(\eta) \le 0\}$ and $\mathscr{P}^{(\mathrm{s})} = \{\boldsymbol{\xi} \in \mathscr{P} \mid \xi - \xi_{\mathscr{S}}(\eta) > 0\}$, respectively. In $\mathscr{P}^{(\mathrm{s})}$, the expression for $\boldsymbol{u}_t^{(\mathrm{s})}(\boldsymbol{\xi})$ in Eq. (3.27) solves Eq. (3.19) with constant sliding direction given by $\hat{\boldsymbol{s}}_t = \boldsymbol{\sigma}/\sigma$[4] and clearly preserves the continuity from adhesion to sliding. The solution (3.27) is valid until the condition on the right-hand side is satisfied, that is if the slope of the total adhesion shear stress $q_t^{(\mathrm{a})}(\xi) = \left\| \boldsymbol{q}_t^{(\mathrm{a})}(\xi) \right\|$ is higher than that of the friction bound. More specifically, the points for which $k\sigma = \mu \partial q_z(\boldsymbol{\xi})/\partial \xi$ are those where the nondimensional sliding velocity $\bar{\boldsymbol{v}}_{\mathrm{s}}(\boldsymbol{\xi})$ vanishes and, thus, correspond to an adhesion edge $\mathscr{A}$. If the pressure distribution is according to Assumption 3.1.2, adhesion is never reestablished from sliding conditions, and thus, the sliding zone is always unique.

[4] When the tyre is anistoropic, a constant sliding direction cannot be found.

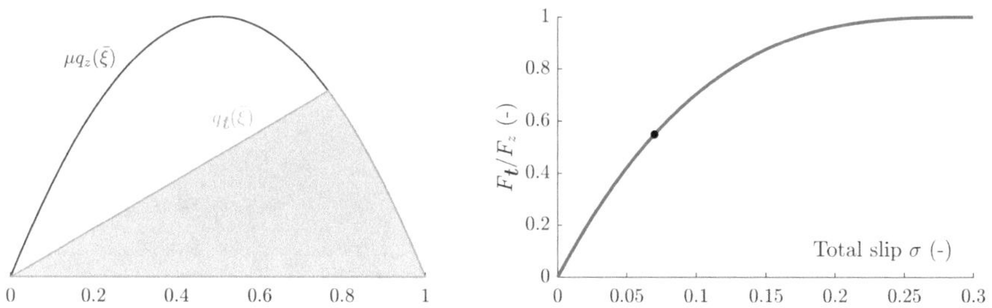

(a) $\sigma = 0.07$.

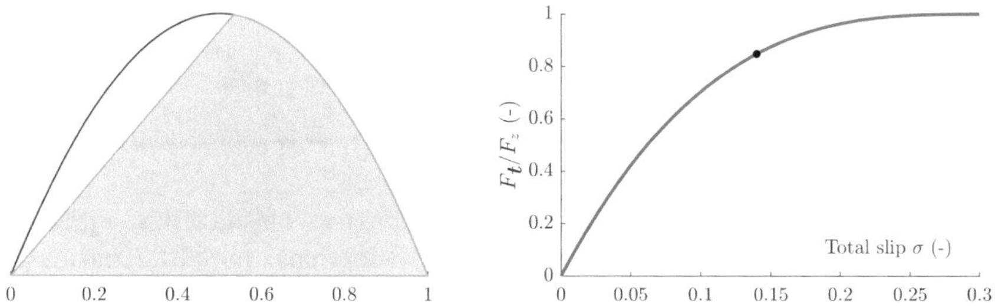

(b) $\sigma = 0.14$.

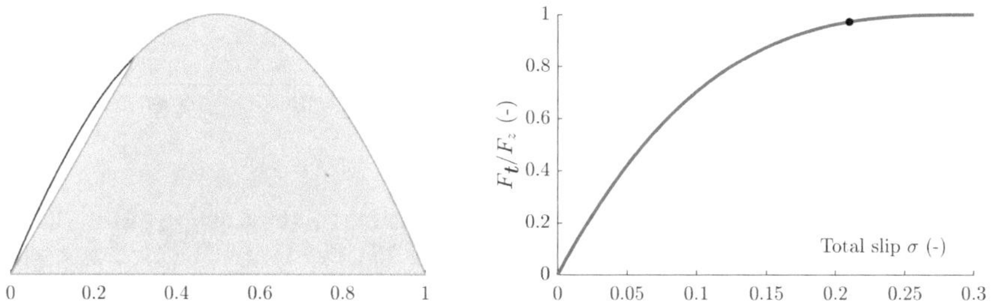

(c) $\sigma = 0.21$.

Fig. 3.1 Shear stress distribution over the contact length for different values of the total slip σ. The corresponding total force amount approximately to 55, 85 and 97% of μF_z

For a rectangular geometry, the distribution of the total shear stress in the contact patch is plotted in Fig. 3.1 versus the nondimensional coordinate $\bar{\xi} = \xi/(2a)$ for three different values of $\sigma = 0.07, 0.14$ and 0.21. On the right-hand side, the attained normalised total forces $F_t(\boldsymbol{\sigma})/F_z$ with $F_t(\boldsymbol{\sigma}) = \left\| \boldsymbol{F}_t(\boldsymbol{\sigma}) \right\|$ are illustrated. These correspond approximately to 55, 85 and 97% of the peak force, which in this model coincides with μF_z.

In particular, the analytical expression for the steady-state forces and moment may again be obtained by integration over $\mathscr{P}$ according to Eqs. (2.24), which yields

$$\boldsymbol{F}_t(\boldsymbol{\sigma}) = C_\sigma \boldsymbol{\sigma}\left[1 - \frac{\sigma}{\sigma^{\mathrm{cr}}} + \frac{1}{3}\left(\frac{\sigma}{\sigma^{\mathrm{cr}}}\right)^2\right], \qquad \sigma < \sigma^{\mathrm{cr}}, \tag{3.31a}$$

$$M_z(\boldsymbol{\sigma}) = -C_{M\sigma_y}\sigma_y\left[1 - 3\frac{\sigma}{\sigma^{\mathrm{cr}}} + 3\left(\frac{\sigma}{\sigma^{\mathrm{cr}}}\right)^2 - \left(\frac{\sigma}{\sigma^{\mathrm{cr}}}\right)^3\right], \qquad \sigma < \sigma^{\mathrm{cr}}, \tag{3.31b}$$

and

$$\boldsymbol{F}_t(\boldsymbol{\sigma}) = \mu F_z\frac{\boldsymbol{\sigma}}{\sigma}, \qquad \sigma \geq \sigma^{\mathrm{cr}}, \tag{3.32a}$$

$$M_z(\boldsymbol{\sigma}) = 0, \qquad \sigma \geq \sigma^{\mathrm{cr}}, \tag{3.32b}$$

with C_σ given as in Eq. (3.15). The corresponding expressions in terms of practical or geometrical slips may be obtained using the relationships in Sect. 1.2, but are not very neat. It may be inferred from Eqs. (3.31a) and (3.32a) that, for an isotropic tyre with pure translational slips, the planar force vector $\boldsymbol{F}_t(\boldsymbol{\sigma})$ is always oriented as the slip. The pneumatic trail is in this case a function of both the translational slips, that is $t_{\mathrm{p}}(\boldsymbol{\sigma})$, and is given by

$$t_{\mathrm{p}}(\boldsymbol{\sigma}) = a\sigma_y\frac{1 - 3\sigma/\sigma^{\mathrm{cr}} + 3\left(\sigma/\sigma^{\mathrm{cr}}\right)^2 - \left(\sigma/\sigma^{\mathrm{cr}}\right)^3}{3 - 3\sigma/\sigma^{\mathrm{cr}} + \left(\sigma/\sigma^{\mathrm{cr}}\right)^2}, \qquad \sigma < \sigma^{\mathrm{cr}}, \tag{3.33}$$

and $t_{\mathrm{p}}(\boldsymbol{\sigma}) = 0$ for $\sigma \geq \sigma^{\mathrm{cr}}$. Once again, the self-aligning moment acting on the virtual contact point may be computed combining Eqs. (2.25), (3.31) and (3.32) clearly with $c_{\mathrm{r}}(0) = 0$.

The tyre forces described by Eqs. (3.31a) and (3.32a) may be interpreted as two surfaces embedded in the three-dimensional space. They are shown in Fig. 3.2 together with the total force $F_t(\boldsymbol{\sigma})$. A similar representation may of course be obtained for the self-aligning moment.

The projections of the tyre characteristics for fixed values of σ_x are plotted in Fig. 3.3. The blue lines reproduce the lateral force and self-aligning moment in pure lateral slip conditions, that is $\sigma_x = 0$. In this case, the lateral force increases up to the critical slip value and then remains constant and equal to μF_z. The moment attains its maximum and minimum values of $\pm(27/256)\mu F_z$ for $\sigma_y = \mp\sigma^{\mathrm{cr}}/4$ and then decays rapidly to zero after the critical slip value. In combined slip conditions, both the lateral force and the moment decrease in absolute value, since part of the available friction is exploited to generate lateral force.

The tyre characteristics $\boldsymbol{F}_t(\boldsymbol{\sigma})$ together with $M_z(\boldsymbol{\sigma})$ may be also thought of as a three-dimensional, regular surface. This is referred to by some authors as the *tyre action surface* [21]. Its projections are traditionally called *Gough plots*. Figure 3.4

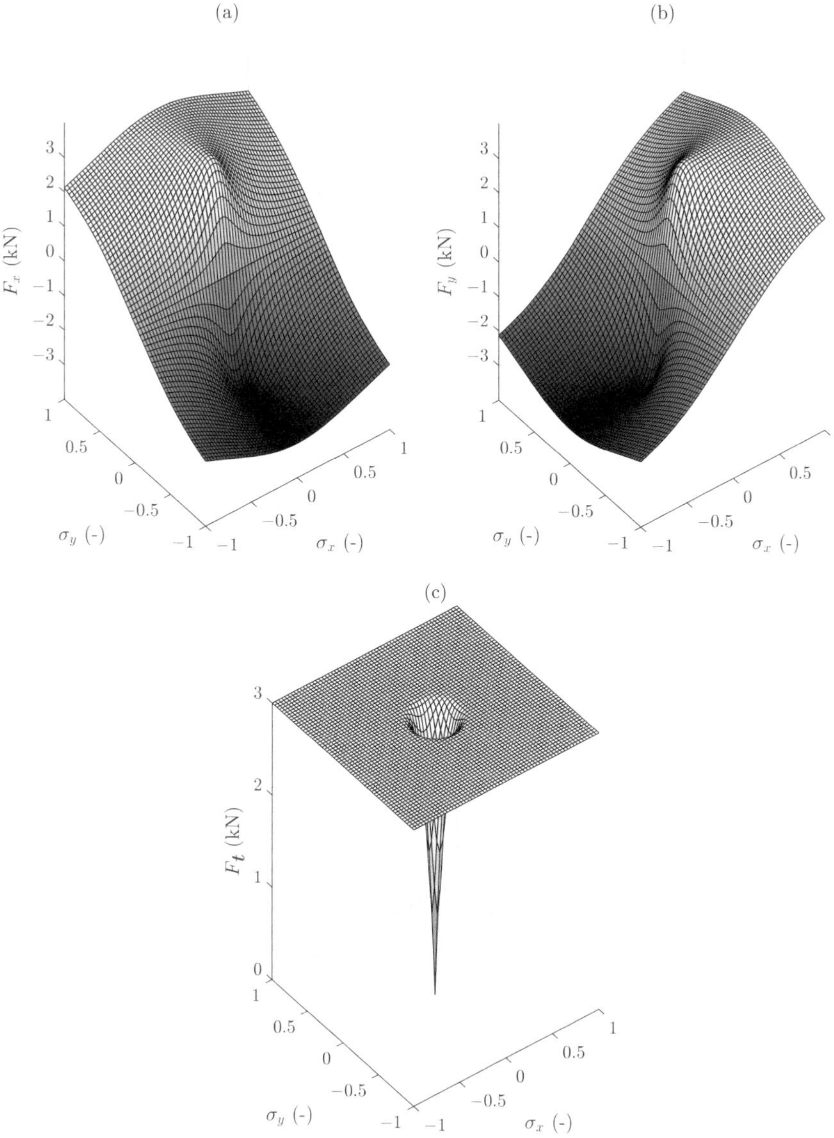

Fig. 3.2 Tyre forces $F_x(\boldsymbol{\sigma})$ and $F_y(\boldsymbol{\sigma})$, together with the total force $F_t(\boldsymbol{\sigma})$, as a function of the theoretical slip variables. The tyre forces may be represented as surfaces embedded in a three-dimensional space

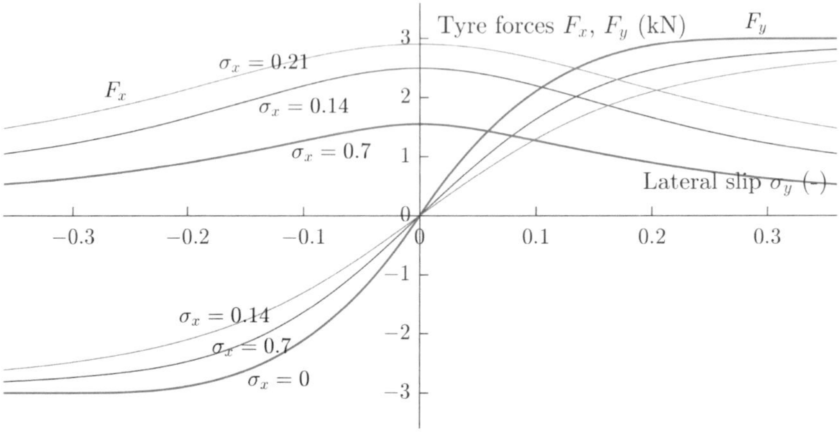

(a) Tyre forces.

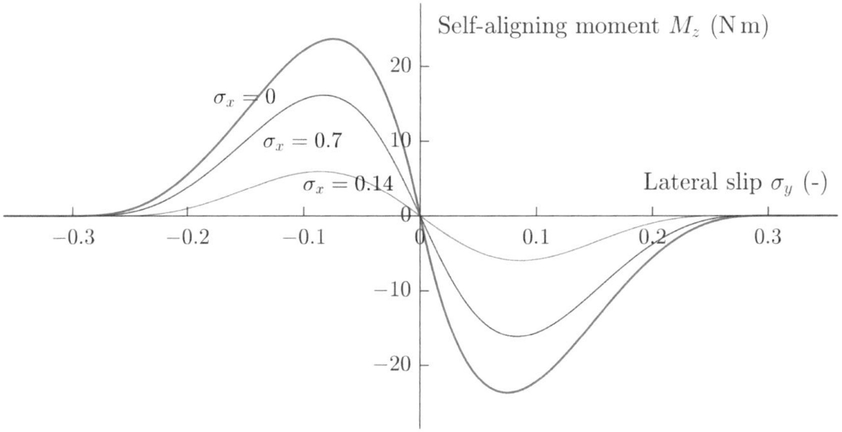

(b) Self-aligning moment.

Fig. 3.3 Tyre characteristics for combined slip conditions

illustrates the Gough plot of the lateral force versus the longitudinal one, for constant values of σ_y. The corresponding lines obtained using the practical slip variables $\boldsymbol{\kappa}$ in place of the theoretical ones are also shown. This kind of representation is often called a friction circle, since, for an isotropic tyre, it encircles all the points for which $F_t \leq \mu F_z$. When the tyre is anisotropic, the friction circle becomes more generally an ellipse.

For an elliptical contact shape, the analytical expression for the tyre characteristics is more involving and not reported here for brevity. The general conclusions that may be drawn are, however, analogous.

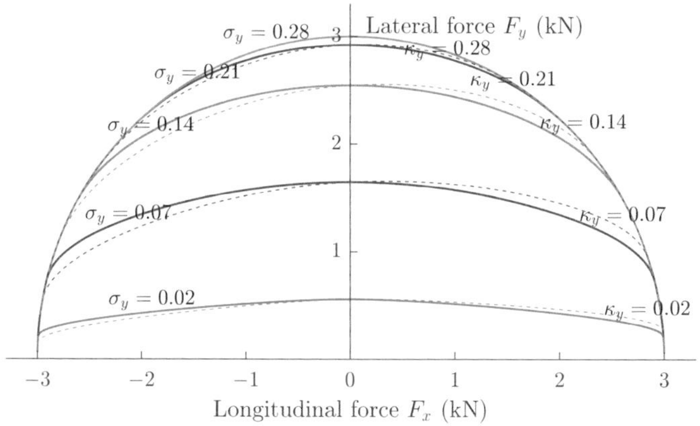

Fig. 3.4 Friction circle

3.4 Pure Spin Slip

The analysis in presence of pure spin slip may be conducted under the assumption of a *thin tyre* [1, 2], which mathematically translates into having the contact patch width sufficiently small, that is $b \ll a$. This allows to approximate the solution inside the whole contact patch by only considering the middle plane of the tyre. More formally, the contact patch may be described in this case as the set $\mathscr{P} = \{\xi \in \mathbb{R} \mid 0 \leq \xi \leq 2a\}$, with the leading edge parametrised by $x = x_{\mathscr{L}} = a$ in the original coordinate frame (the dependency on y or η is henceforth omitted for the sake of notation).

The following analysis considers again a tyre with isotropic tread and diagonal stiffness matrix. The pressure distribution is assumed to be parabolic as in Eq. (3.22). In this context, a crucial role is played by the so-called critical spin φ^{cr}, defined as

$$\varphi^{\text{cr}} \equiv \frac{\sigma^{\text{cr}}}{a} \triangleq 2\mu \frac{q_z^*}{ka^2} = \frac{\mu F_z}{C_\varphi}, \tag{3.34}$$

in which the spin stiffness C_φ reads as in Eq. (3.15). It may be easily verified that, for non-supercritical values of the spin slip, that is $|\varphi| \leq \varphi^{\text{cr}}$, sliding never occurs in the contact patch ($\mathscr{P}^{(\text{a})} \equiv \mathscr{P}$). Therefore, the global adhesion solution $\boldsymbol{u}_t^{(\text{a})}(\xi) \in C^\infty(\mathscr{P}; \{0\} \times \mathbb{R})$ may be assumed to be of the form $\boldsymbol{u}_t^{(\text{a})}(\xi) = u_y^{(\text{a})}(\xi)\hat{\boldsymbol{e}}_y$, where the lateral deflection is obviously given by Eq. (3.13), with $\boldsymbol{\sigma} = \mathbf{0}$. For the present case, it reads specifically

$$u_y^{(\text{a})}(\xi) = \frac{1}{2}\varphi\xi(2a - \xi), \qquad \xi \in \mathscr{P}. \tag{3.35}$$

The analytical expression for the lateral force and moment acting on the tyre may be calculated as follows[5]:

$$F_y(\varphi) = C_{\varphi}\varphi, \qquad \left|\varphi\right| \leq \varphi^{\mathrm{cr}}, \tag{3.36a}$$

$$M_z(\varphi) = 0, \qquad \left|\varphi\right| \leq \varphi^{\mathrm{cr}}. \tag{3.36b}$$

On the other hand, for supercritical values of the spin $\left|\varphi\right| > \varphi^{\mathrm{cr}}$, two sliding zones arise in the contact patch, which are located at the leading and trailing edges, respectively. In this case, the lateral component of the sliding solution may be generally sought in the form

$$u_y^{(\mathrm{s})}(\xi) = \frac{\mu}{k}q_z(\xi)\,\mathrm{sgn}\,\varphi \qquad \text{if} \qquad \left(\left|\varphi\right| - \varphi^{\mathrm{cr}}\right)(a-\xi) > 0, \tag{3.37a}$$

$$u_y^{(\mathrm{s})}(\xi) = -\frac{\mu}{k}q_z(\xi)\,\mathrm{sgn}\,\varphi \qquad \text{if} \qquad \left(\left|\varphi\right| + \varphi^{\mathrm{cr}}\right)(a-\xi) < 0, \tag{3.37b}$$

where the conditions on the right-hand sides ensure nonzero values of the micro-sliding speed. Accordingly, the constant sliding directions in Eq. (3.19) are given, respectively, by $\hat{\boldsymbol{s}}_t = \pm\,\mathrm{sgn}\,\varphi\hat{\boldsymbol{e}}_y$.

In particular, the bristles start sliding from the leading edge, which coincides with a first sliding edge $\mathscr{S}_1 = \{\xi \in \mathscr{P} \mid \xi = 0\}$ (since $\xi_{\mathscr{S}_1} = 0$). On $\mathscr{S}_1$ the sign of the deflection must be concordant with that of the spin slip. Therefore, a particular solution to (3.19) satisfying the BC (2.23) is given by $\boldsymbol{u}_t^{(\mathrm{s})}(\xi) = u_y^{(\mathrm{s})}(\xi)\hat{\boldsymbol{e}}_y$, with $u_y^{(\mathrm{s})}(\xi)$ as in Eq. (3.37a). Sliding continues until the condition on the right-hand side of Eq. (3.37a) is satisfied, that is in $\mathscr{P}_1^{(\mathrm{s})} = \{\xi \in \mathscr{P} \mid \xi < a\}$. Adhesion is then reestablished at $\xi = \xi_{\mathscr{A}} = a$, which parametrise the adhesion edge $\mathscr{A} = \{\xi \in \mathscr{P} \mid \xi - a = 0\}$. In this point, the slope of the friction and camber parabola is equal. The adhesion solution may be calculated starting from Eq. (3.11) with the following BC:

$$\text{BC:} \qquad \boldsymbol{u}_t(a) = \frac{\mu}{k}q_z^*\,\mathrm{sgn}\,\varphi\hat{\boldsymbol{e}}_y, \tag{3.38}$$

which preserves the continuity in the transition from sliding to adhesion conditions. Enforcing the above BC (3.38) yields the following expression for the lateral adhesion solution in $\mathscr{P}^{(\mathrm{a})} = \{\xi \in \mathscr{P} \mid a \leq \xi \leq \xi_{\mathscr{S}_2}\}$:

$$u_y^{(\mathrm{a})}(\xi) = -\frac{1}{2}\varphi(a-\xi)^2 + \frac{1}{2}\varphi^{\mathrm{cr}}a^2\,\mathrm{sgn}\,\varphi, \qquad \xi \in \mathscr{P}^{(\mathrm{a})}. \tag{3.39}$$

[5] It should be noticed that, since it is assumed that the contact patch has no lateral dimension, the analytical expressions for the stiffnesses in Eq. (3.15) should be reduced by a factor of $2b$. This consideration also holds for Sect. 3.5.

In the definition of the adhesion zone $\mathscr{P}^{(a)}$, $\xi_{\mathscr{S}_2}$ parametrises the coordinate of the second sliding edge $\mathscr{S}_2$, that is point where the lateral shear stress $q_y^{(a)}(\xi) = ku_y^{(a)}(\xi)$ computed starting from Eq. (3.39) intercepts the friction parabola that is discordant with the spin slip. This happens for $\xi = \xi_{\mathscr{S}_2}$, with

$$\xi_{\mathscr{S}_2} = a\left(1 + \frac{\sqrt{2}}{\sqrt{1+|\varphi|/\varphi^{\mathrm{cr}}}}\right), \qquad |\varphi| > \varphi^{\mathrm{cr}}, \tag{3.40}$$

and therefore, the second sliding zone, where the solution to Eq. (3.19) is given by Eq. (3.37b), may be defined as $\mathscr{P}_2^{(s)} = \{\xi \in \mathscr{P} \mid \xi_{\mathscr{S}_2} < \xi \le 2a\}$. The global solution $\boldsymbol{u}_t(\xi) \in C^0(\mathscr{P}; \{0\} \times \mathbb{R})$ may be then constructed as

$$\boldsymbol{u}_t(\xi) = \begin{cases} \boldsymbol{u}_t^{(a)}(\xi) = u_y^{(a)}(\xi)\hat{\boldsymbol{e}}_y, & \xi \in \mathscr{P}^{(a)}, \\ \boldsymbol{u}_t^{(s)}(\xi) = u_y^{(s)}(\xi)\hat{\boldsymbol{e}}_y, & \xi \in \mathscr{P}^{(s)}, \end{cases} \tag{3.41}$$

where $u_y^{(a)}(\xi)$ is as in Eq. (3.39), $\mathscr{P}^{(s)} = \mathscr{P}_1^{(s)} \cup \mathscr{P}_2^{(s)}$ and

$$u_y^{(s)}(\xi) = \begin{cases} \dfrac{\mu}{k} q_z(\xi)\,\mathrm{sgn}\,\varphi, & \xi \in \mathscr{P}_1^{(s)}, \\ -\dfrac{\mu}{k} q_z(\xi)\,\mathrm{sgn}\,\varphi, & \xi \in \mathscr{P}_2^{(s)}. \end{cases} \tag{3.42}$$

The distribution of the lateral shear stress due to pure spin conditions is illustrated in Fig. 3.5 for different values of $\varphi = 3, 15$ and $30\,\mathrm{m}^{-1}$, respectively. The corresponding values attained by the lateral force are shown on the right-hand side and amount approximately to the 75, 66 and 49% of the peak force, again given by μF_z. In particular, the first shear stress distribution in Fig. 3.5 results from subcritical spin slip conditions ($\varphi^{\mathrm{cr}} = 4.00\,\mathrm{m}^{-1}$ in figure), whilst the other two plots illustrate the trends obtained for supercritical spin slips and are characterised by two sliding zones separated by an adhesion region.

The lateral force and moment acting on the tyre in pure supercritical spin slip conditions may be finally obtained by integration:

$$F_y(\varphi) = \sqrt{2}\mu F_z \frac{\mathrm{sgn}\,\varphi}{\sqrt{1+|\varphi|/\varphi^{\mathrm{cr}}}}, \qquad |\varphi| > \varphi^{\mathrm{cr}}, \tag{3.43a}$$

$$M_z(\varphi) = \frac{3}{8}\mu F_z a \frac{|\varphi|/\varphi^{\mathrm{cr}} - 1}{|\varphi|/\varphi^{\mathrm{cr}} + 1}\,\mathrm{sgn}\,\varphi, \qquad |\varphi| > \varphi^{\mathrm{cr}}. \tag{3.43b}$$

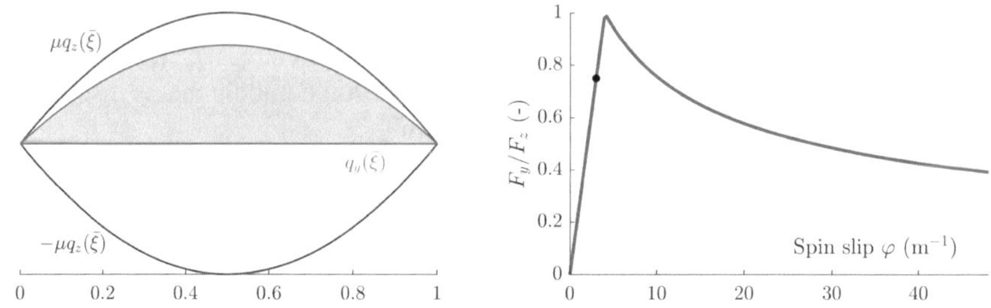

(a) φ = 3.

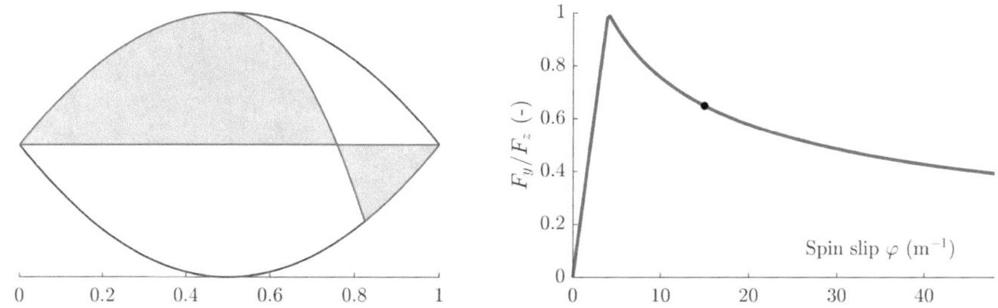

(b) φ = 15.

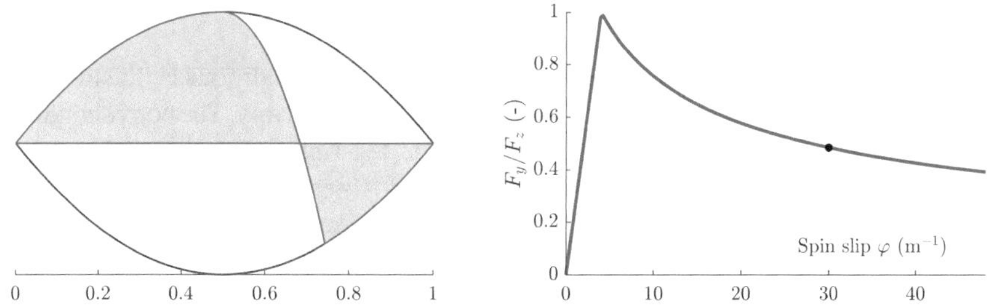

(c) φ = 30.

Fig. 3.5 Lateral stress distribution over the contact length for different values of the spin slip φ. The corresponding total force amount approximately to 75, 66 and 49% of μF_z

Both the tyre characteristics are reproduced graphically in Fig. 3.6. It may be observed that the lateral force is linear for $|\varphi| \leq \varphi^{\text{cr}}$ and then starts slowly to decay to zero. On the other hand, the self-aligning moment is only excited for supercritical values of the spin slips and tends to opposite asymptotes $\pm(3/8)\mu F_z a$. Both functions are continuous but not differentiable at $|\varphi| = \varphi^{\text{cr}}$.

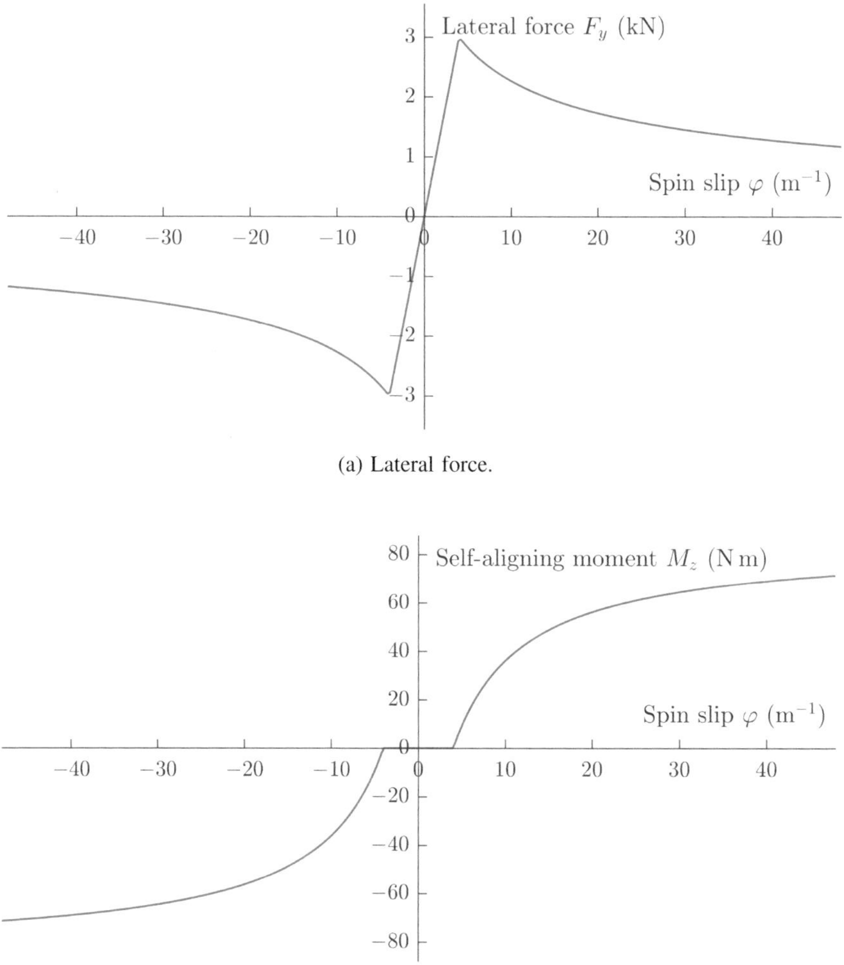

(a) Lateral force.

(b) Self-aligning moment.

Fig. 3.6 Lateral force and self-aligning moment for pure spin slip conditions

3.5 Lateral Slip and Spin

The case of combined lateral and spin slip is more involving than the previous ones. In particular, the analysis proposed in this book restricts to a thin tyre, subcritical values of the spin parameter $|\varphi| < \varphi^{\text{cr}}$ and parabolic pressure distribution $q_z(\xi)$ as in Eq. (3.22). For a complete treatment of the case $|\varphi| \geq \varphi^{\text{cr}}$, the reader is referred to [1, 3].

With respect to the pure lateral problem, the effect of a subcritical spin is to increase or reduce the friction coefficient available inside the contact patch [4]. To understand this aspect, it may be convenient to introduce

$$\tilde{\mu} \triangleq \mu\left[1 - \frac{|\varphi|}{\varphi^{\text{cr}}}\operatorname{sgn}(\sigma_y\varphi)\right], \qquad |\varphi| < \varphi^{\text{cr}}, \tag{3.44}$$

and consider an equivalent friction parabola $\tilde{\mu}q_z(\xi)$. Accordingly, the equivalent critical slip that causes full sliding inside the contact patch may be defined as

$$\tilde{\sigma}^{\text{cr}} \triangleq \sigma^{\text{cr}} - a|\varphi|\operatorname{sgn}(\sigma_y\varphi) = a\varphi^{\text{cr}} - a|\varphi|\operatorname{sgn}(\sigma_y\varphi), \qquad |\varphi| < \varphi^{\text{cr}}. \tag{3.45}$$

Actually, it may be observed that the quantity $\tilde{\sigma}^{\text{cr}}$ in Eq. (3.45) coincides with the definition of critical slip given in Eq. (3.21), provided that the friction coefficient μ is replaced by $\tilde{\mu}$:

$$\tilde{\sigma}^{\text{cr}} \equiv \frac{\tilde{\mu}}{k}\left|\frac{\partial q_z(\xi)}{\partial \xi}\right|\Bigg|_{\xi=0} = 2\tilde{\mu}\frac{q_z^*}{ka} = \frac{3\tilde{\mu}F_z}{C_\sigma}. \tag{3.46}$$

Therefore, the analysis may be conducted similarly as done in Sect. 3.3. However, in this case, a global solution should be sought in the form $\boldsymbol{u}_t(\xi) = u_y(\xi)\hat{\boldsymbol{e}}_y$. In particular, for values of $|\sigma_y| < \tilde{\sigma}^{\text{cr}}$, an adhesion zone originates at the trailing edge. The expression for the lateral displacement may be deduced from Eqs. (3.13) with $\sigma_x = 0$ and reads

$$u_y^{(\text{a})}(\xi) = \sigma_y\xi + \frac{1}{2}\varphi\xi(2a - \xi), \qquad \xi \in \mathscr{P}^{(\text{a})}. \tag{3.47}$$

The adhesion solution above is valid until the lateral shear stress $q_y^{(\text{a})}(\xi) = ku_y^{(\text{a})}(\xi)$ intercepts the friction parabola that is concordant with the sign of the lateral slip σ_y. Specifically, this happens for $\xi = \xi_{\mathscr{S}}$, with

$$\xi_{\mathscr{S}} = 2a\left(1 - \frac{|\sigma_y|}{\tilde{\sigma}^{\text{cr}}}\right), \qquad |\sigma_y| \le \tilde{\sigma}^{\text{cr}}. \tag{3.48}$$

In the sliding zone $\mathscr{P}^{(\text{s})}$, the lateral sliding solution may be sought in the form

$$u_y^{(\text{s})}(\xi) = \frac{\mu}{k}q_z(\xi)\operatorname{sgn}\sigma_y \quad \text{if} \quad |\sigma_y| > \Big(\varphi^{\text{cr}} - |\varphi|\operatorname{sgn}(\sigma_y\varphi)\Big)(a - \xi), \tag{3.49}$$

with constant sliding direction given by $\hat{\boldsymbol{s}}_t = \operatorname{sgn}\sigma_y\hat{\boldsymbol{e}}_y$. The expression for $u_y^{(\text{s})}(\xi)$ as in Eq. (3.49) satisfies the BC (2.23) on $\mathscr{S} = \{\xi \in \mathscr{P} \mid \xi - \xi_{\mathscr{S}} = 0\}$ and represents a particular solution to (3.19). It may be noticed that the condition on the right-hand side of Eq. (3.49) may be rearranged as

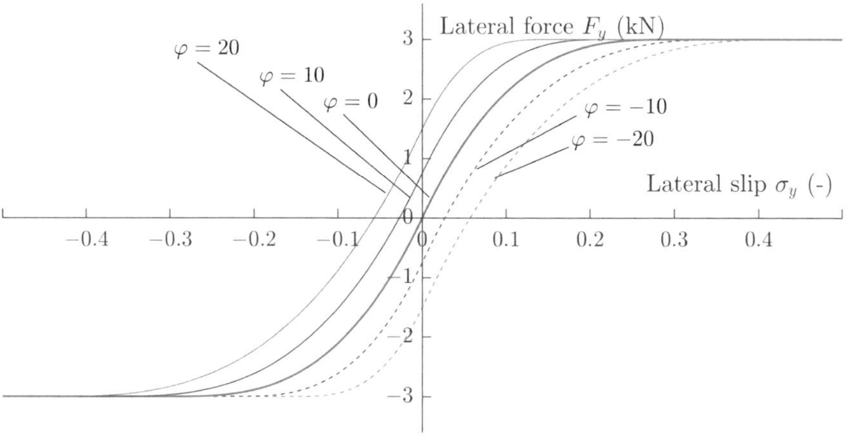

(a) Lateral force.

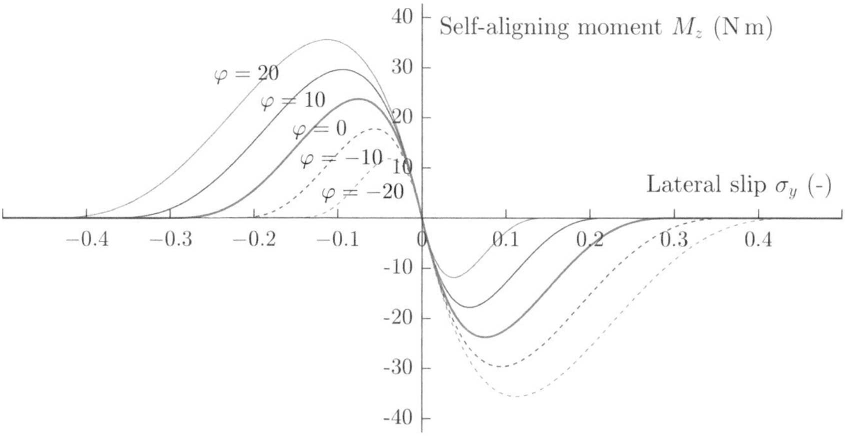

(b) Self-aligning moment.

Fig. 3.7 Lateral force and self-aligning moment for combined lateral and spin slip conditions

$$k\left|\sigma_y\right| > \tilde{\mu}\,\frac{\partial q_z(\xi)}{\partial \xi}\,, \tag{3.50}$$

which is analogous to the condition for the case of pure translational slip, again with μ replaced by $\tilde{\mu}$. This means that the adhesion may be reestablished only where the slope of the lateral shear stress due to pure slip $q_{y\alpha} \triangleq k\sigma_y$ equals that of the equivalent friction parabola. Such a point, if exists (actually only for $\left|\sigma_y\right| \leq \tilde{\sigma}^{\mathrm{cr}}$), is never located after $\xi = \xi_{\mathscr{S}}$ (Lemma 4.A.8). Clearly, when $\left|\sigma_y\right| \geq \tilde{\sigma}^{\mathrm{cr}}$, full sliding occurs inside the contact patch and only the sliding solution is retained.

More generally, defining $\mathscr{P}^{(a)} \triangleq \{\xi \in \mathscr{P} \mid 0 \leq \xi \leq \xi_{\mathscr{S}}\}$ and $\mathscr{P}^{(s)} \triangleq \{\xi \in \mathscr{P} \mid \xi_{\mathscr{S}} < \xi \leq 2a\}$, with $\xi_{\mathscr{S}}$ as in Eq. (3.48), the global solution $\boldsymbol{u}_t(\xi) \in C^0(\mathscr{P}; \{0\} \times \mathbb{R})$ may be formally constructed as in Eq. (3.41), but with $u_y^{(a)}(\xi)$ and $u_y^{(s)}(\xi)$ reading as in Eqs. (3.47) and (3.49), respectively. Indeed, it is possible to show that this definition of the adhesion and sliding zone is consistent with the original one given by Eqs. (2.4) (Lemma 4.A.8).

Integrating over the contact patch yields closed-form expression for the lateral force and the self-aligning torque:

$$F_y(\sigma_y, \varphi) = C_\sigma \sigma_y \left[1 - \left|\frac{\sigma_y}{\tilde{\sigma}^{cr}}\right| + \frac{1}{3}\left(\frac{\sigma_y}{\tilde{\sigma}^{cr}}\right)^2\right] + C_\varphi \varphi, \qquad |\sigma_y| < \tilde{\sigma}^{cr}, \tag{3.51a}$$

$$M_z(\sigma_y, \varphi) = -C_{M\sigma_y} \sigma_y \left[1 - 3\left|\frac{\sigma_y}{\tilde{\sigma}^{cr}}\right| + 3\left(\frac{\sigma_y}{\tilde{\sigma}^{cr}}\right)^2 - \left|\frac{\sigma_y}{\tilde{\sigma}^{cr}}\right|^3\right], \quad |\sigma_y| < \tilde{\sigma}^{cr}, \tag{3.51b}$$

and

$$F_y(\sigma_y, \varphi) = \mu F_z \operatorname{sgn} \sigma_y, \qquad |\sigma_y| \geq \tilde{\sigma}^{cr}, \tag{3.52a}$$

$$M_z(\sigma_y, \varphi) = 0, \qquad |\sigma_y| \geq \tilde{\sigma}^{cr}. \tag{3.52b}$$

Both quantities are depicted in Fig. 3.7 for different values of the spin slip φ.

References

1. Pacejka HB (2012) Tire and vehicle dynamics, 3rd edn. Elsevier/BH, Amsterdam
2. Limebeer DJN, Massaro M (2018) Dynamics and optimal control of road vehicle. Oxford University Press
3. Sakai H (1990) Study on cornering properties of tire and vehicle. Tire Sci Technol 18(3):136–169. Available from: https://doi.org/10.2346/1.2141697
4. Svendenius J (2007) Tire modelling and friction estimation [dissertation]. Lund
5. Pacejka HB, Bakker E (1992) The magic formula tire model. Int J Vehic Mech Mob 21(1):1–18
6. Bayle P, Forissier JF, Lafon S (1993) A new tyre model for vehicle dynamics simulations. Autom Technol Int, 193–198
7. Pacejka HB, Besselink IJM (1997) Magic formula tyre model with transient properties. Vehicle Syst Dyn 27(sup001):234–249
8. Besselink IJM, Schmeitz AJC, Pacejka HB (2010) An improved magic formula/swift tyre model that can handle inflation pressure changes. Vehicle Syst Dyn 48(sup1):337–352
9. Farroni F, Giordano D, Russo M et al. (2014) TRT: thermo racing tyre a physical model to predict the tyre temperature distribution. Meccanica 49:707–723. Available from: https://doi.org/10.1007/s11012-013-9821-9
10. Farroni F (2016) T.R.I.C.K.-tire/Road interaction characterization & knowledge—a tool for the evaluation of tire and vehicle performances in outdoor test sessions. Mechan Syst Signal

Proc 72–73:808-831. ISSN 0888-3270. Available from: https://doi.org/10.1016/j.ymssp.2015.11.019
11. Farroni F, Russo M, Sakhnevych A, Timpone F (2019) TRT EVO: Advances in real-time thermodynamic tire modeling for vehicle dynamics simulations. Proc IMechE Part D: J Autom Eng 233(1):121–135
12. Farroni F, Sakhnevych A, Timpone F (2017) Physical modelling of tire wear for the analysis of the influence of thermal and frictional effects on vehicle performance. Proc Inst Mech Eng Part L: J Mater Des Appl 231(1–2):151–161
13. Sakhnevych A (2021) Multiphysical MF-based tyre modelling and parametrisation for vehicle setup and control strategies optimisation. Vehicle Syst. Dyn. 2021. Available from: https://doi.org/10.1080/00423114.2021.1977833
14. Gipser M (2007) FTire—the tire simulation model for all applications related to vehicle dynamics. Vehicle Syst Dyn 45(S1):139–151. Available from: https://doi.org/10.1080/00423110801899960
15. Gipser M (2016) FTire and puzzling tyre physics: teacher, not student. Vehicle Syst Dyn 54(4):448–462. Available from: https://doi.org/10.1080/00423114.2015.1117116
16. Gipser M (2005) FTire: a physically based application-oriented tyre model for use with detailed MBS and finite-element suspension models. Vehicle Syst Dyn 43(sup1):76–91. Available from: https://doi.org/10.1080/00423110500139940
17. Available from: https://www.itwm.fraunhofer.de/en/departments/mf/cdtire.html
18. Romano L, Timpone F, Bruzelius F, Jacobson B. Analytical results in transient brush tyre models: theory for large camber angles and classic solutions with limited friction. Meccanica. Available from: https://doi.org/10.1007/s11012-021-01422-3
19. Matoušek, J (2001) On directional convexity. Discrete Comput Geom 25:389–403. Available from: https://doi.org/10.1007/s004540010069
20. Evans LC (2010) Partial differential equations. 2nd ed. American Mathematical Society
21. Guiggiani M (2018) The science of vehicle dynamics, 2nd edn. Springer International, Cham(Switzerland)

Chapter 4
Unsteady-State Brush Theory

Abstract Many transient phenomena concerning the tyre-road interaction are effectively explained within the theoretical framework of the brush theory. The analysis in vanishing sliding conditions is relatively simple and may be conducted with respect to any time-varying slip input. The case of limited friction available inside the contact patch is rather involving. In this context, the investigations proposed in this chapter are limited to small spin slips under the assumption of a thin tyre. The situation further complicates when considering a flexible tyre carcass, but it may be still approached using some intuition from Chap. 3. A rather general formulation of the transient problem is proposed, which allows to gain some preliminary insights about the relaxation behaviour of the tyre.

Traditionally, the transient response of the pneumatic tyre subjected to variable slip inputs is studied using more sophisticated formulations than the brush theory introduced in Chap. 2 . These include, for example, the famous *stretched string models* [1–11]. Other proposed approaches rely on numerical techniques [12, 13]. However, many phenomena may still be explained within the simpler theoretical framework provided by the brush theory, at least qualitatively.

The scope of the present chapter is not that of deriving a set of relationships between the tyre characteristics and the slip variables, but to acquiring a basic understanding of the nonstationary behaviour of the tyre. These insights will reveal to be extremely useful in the development of pragmatic models for vehicle dynamics simulations. This aspect will be discussed more extensively in Chap. 6.

The transient problem is first studied under the assumptions of vanishing sliding and rigid tyre carcass in Sect. 4.1. Then, transient solutions are provided for the case of limited friction in Sect. 4.2 with reference to all the one-dimensional problems. These investigations are limited to sufficiently small spin slips φ. Finally, in Sect. 4.3 the dynamics of the tyre carcass is considered. Throughout all the analysis, Assumptions 3.1.1 and 3.1.2 are retained.

L. Romano, *Advanced Brush Tyre Modelling*,
SpringerBriefs in Applied Sciences and Technology,
https://doi.org/10.1007/978-3-030-98435-9_4

As for the steady-state case with fixed contact shape, the problem may be better described replacing the original space coordinates $\boldsymbol{x}$ with the local ones $\boldsymbol{\xi}$ defined as in Eqs. (3.5). Accordingly, Eq. (2.15) may be restated more conveniently as

$$\bar{\boldsymbol{v}}_{\mathrm{s}}(\boldsymbol{\xi}, s) = -\boldsymbol{\sigma}(s) - \mathbf{A}_{\varphi}(s) \begin{bmatrix} x_{\mathscr{L}}(\eta) - \xi \\ \eta \end{bmatrix} + \frac{\partial \boldsymbol{u}_t(\boldsymbol{\xi}, s)}{\partial s} + \frac{\partial \boldsymbol{u}_t(\boldsymbol{\xi}, s)}{\partial \xi}, \quad (\boldsymbol{\xi}, s) \in \mathring{\mathscr{P}} \times \mathbb{R}_{>0}. \tag{4.1}$$

4.1 Vanishing Sliding

Assuming vanishing sliding conditions, that is $\bar{\boldsymbol{v}}_{\mathrm{s}}(\boldsymbol{\xi}, s) = \mathbf{0}$ all over the contact patch ($\mathscr{P}^{(\mathrm{a})} \equiv \mathscr{P}$), Eq. (4.1) becomes

$$\frac{\partial \boldsymbol{u}_t(\boldsymbol{\xi}, s)}{\partial s} + \frac{\partial \boldsymbol{u}_t(\boldsymbol{\xi}, s)}{\partial \xi} = \boldsymbol{\sigma}(s) + \mathbf{A}_{\varphi}(s) \begin{bmatrix} x_{\mathscr{L}}(\eta) - \xi \\ \eta \end{bmatrix}, \quad (\boldsymbol{\xi}, s) \in \mathring{\mathscr{P}} \times \mathbb{R}_{>0}. \tag{4.2}$$

Equation (4.2) represents the nonstationary version of Eq. (3.10) in Chap. 3. More specifically, it consists of two linear, uncoupled transport PDEs involving only two partial derivatives: one taken with respect to the longitudinal coordinate ξ and another with respect to the travelled distance s. In order to be solved, they need to be supplemented with a proper BC and an IC. When the contact patch is fixed, that is $\mathscr{P}(s) \equiv \mathscr{P}(0) = \mathscr{P}_0$, and recalling that the change of variables from $\boldsymbol{x} \mapsto \boldsymbol{\xi}$ straightens the leading edge $\mathscr{L}$, these may be formulated respectively as follows[1]:

$$\text{BC:} \qquad \boldsymbol{u}_t(0, \eta, s) = \mathbf{0}, \qquad s \in \mathbb{R}_{>0}, \tag{4.3}$$

$$\text{IC:} \qquad \boldsymbol{u}_t(\boldsymbol{\xi}, 0) = \boldsymbol{u}_{t0}(\boldsymbol{\xi}), \qquad \boldsymbol{\xi} \in \mathring{\mathscr{P}}. \tag{4.4}$$

If the contact patch satisfies Assumption 3.1.1, enforcing the BC (4.3) and IC (4.4), in turn, provides two different solutions to the PDE (4.2). These solutions are uniquely defined on $\mathscr{P}$, and may be sought using the method of the characteristics lines, which yields

[1] It should be observed that if the initial conditions are described by a function $\boldsymbol{u}_{t0}(\boldsymbol{x})$ using the original coordinates $\boldsymbol{x}$, then in the coordinate system $\boldsymbol{\xi}$ they should be represented by a new function $\boldsymbol{u}'_{t0}(\boldsymbol{\xi}) \triangleq \boldsymbol{u}_{t0}(\boldsymbol{x}(\boldsymbol{\xi}))$. Similar considerations also hold for other functions. However, the same notation is used in the remaining of the chapter for the sake of simplicity.

$$\boldsymbol{u}_t^-(\boldsymbol{\xi}, s) = \int_0^{\xi} \boldsymbol{\sigma}(\xi' - \xi + s) + \mathbf{A}_\varphi(\xi' - \xi + s)\begin{bmatrix} x_{\mathscr{L}}(\eta) - \xi' \\ \eta \end{bmatrix} \mathrm{d}\xi', \qquad (\boldsymbol{\xi}, s) \in \mathscr{P}^- \times \mathbb{R}_{\geq 0}, \tag{4.5a}$$

$$\boldsymbol{u}_t^+(\boldsymbol{\xi}, s) = \int_0^{s} \boldsymbol{\sigma}(s') + \mathbf{A}_\varphi(s')\begin{bmatrix} x_{\mathscr{L}}(\eta) - s' + s - \xi \\ \eta \end{bmatrix} \mathrm{d}s' + \boldsymbol{u}_{t0}(\xi - s, \eta), \quad (\boldsymbol{\xi}, s) \in \mathscr{P}^+ \times \mathbb{R}_{\geq 0}. \tag{4.5b}$$

In Eqs. (4.5), the deflections $\boldsymbol{u}_t^-(\boldsymbol{\xi}, s)$ and $\boldsymbol{u}_t^+(\boldsymbol{\xi}, s)$ result from the application of the BC and IC, in turn, respectively. The corresponding subdomains $\mathscr{P}^-$ and $\mathscr{P}^+$ of the contact patch may be defined by introducing the function

$$\gamma_\Sigma(\boldsymbol{\xi}, s) \triangleq \xi - s \tag{4.6}$$

and setting

$$\mathscr{P}^- \triangleq \{\boldsymbol{\xi} \in \mathscr{P} \mid \gamma_\Sigma(\boldsymbol{\xi}, s) < 0\}, \tag{4.7a}$$

$$\mathscr{P}^+ \triangleq \{\boldsymbol{\xi} \in \mathscr{P} \mid \gamma_\Sigma(\boldsymbol{\xi}, s) \geq 0\}. \tag{4.7b}$$

The curve described implicitly by

$$\gamma_\Sigma(\boldsymbol{\xi}, s) = 0, \qquad (\boldsymbol{\xi}, s) \in \mathscr{P} \times \mathbb{R}_{\geq 0} \tag{4.8}$$

or explicitly by

$$\xi = \xi_\Sigma(s) = s, \qquad (\xi, s) \in \big[0, \xi_{\mathscr{T}}(\eta)\big] \times \mathbb{R}_{\geq 0} \tag{4.9}$$

is called a *travelling edge*, because it travels with the rolling speed of the tyre and separates the two solutions $\boldsymbol{u}_t^-(\boldsymbol{\xi}, s)$ and $\boldsymbol{u}_t^+(\boldsymbol{\xi}, s)$. For a rectangular contact patch, $\gamma_\Sigma(\boldsymbol{\xi}, s) = 0$ is a straight line parallel to the η-axis, and coincides with the contact geometry translated by $-s$ in the longitudinal direction. Similarly, for an elliptical patch, translating the contact ellipse by $-s$, the curve $\gamma_\Sigma(\boldsymbol{\xi}, s) = 0$ intersects the trailing edge in $(s, \pm 2b\sqrt{1 - (s/(2a))^2})$.

The global solution $\boldsymbol{u}_t(\boldsymbol{\xi}, s)$ over $\mathscr{P}$ may then be constructed as

$$\boldsymbol{u}_t(\boldsymbol{\xi}, s) = \begin{cases} \boldsymbol{u}_t^-(\boldsymbol{\xi}, s), & (\boldsymbol{\xi}, s) \in \mathscr{P}^- \times \mathbb{R}_{\geq 0}, \\ \boldsymbol{u}_t^+(\boldsymbol{\xi}, s), & (\boldsymbol{\xi}, s) \in \mathscr{P}^+ \times \mathbb{R}_{\geq 0}, \end{cases} \tag{4.10}$$

since $\mathscr{P} = \mathscr{P}^- \cup \mathscr{P}^+$. It may be easily observed that $\boldsymbol{u}_t(\boldsymbol{\xi}, s) \in C^0(\mathscr{P} \times \mathbb{R}_{\geq 0}; \mathbb{R}^2)$ since the deflections $\boldsymbol{u}_t^-(\boldsymbol{\xi}, s)$ and $\boldsymbol{u}_t^+(\boldsymbol{\xi}, s)$ are continuous on the travelling edge $\xi = s$, that is $\boldsymbol{u}_t^-(s, \eta, s) \equiv \boldsymbol{u}_t^+(s, \eta, s)$. This is a direct consequence of the fact that it must necessarily be $\boldsymbol{u}_{t0}(0, \eta) = \mathbf{0}$ from frictional considerations. Unfortunately, the continuity at $\xi = s$ is the only requirement that the global solution may be expected to fulfil. This consideration holds also true for the solution $\boldsymbol{u}_t^+(\boldsymbol{\xi}, s)$, which is often only $C^0(\mathscr{P}^+ \times \mathbb{R}_{\geq 0}; \mathbb{R}^2)$, unless $\boldsymbol{u}_{t0}(\boldsymbol{\xi}) \in C^1(\mathscr{P}; \mathbb{R}^2)$. Actually, this only happens if the initial condition itself corresponds to a distribution which results already from a stationary configuration in vanishing sliding conditions.[2]

This aspect is perhaps better understood by observing that the expressions $\boldsymbol{u}_t^-(\boldsymbol{\xi}, s)$ and $\boldsymbol{u}_t^+(\boldsymbol{\xi}, s)$ in Eqs. (4.5) may be interpreted as the stationary and the transient solutions to the PDEs (4.2), respectively. Indeed, it may be easily inferred from the definition of $\mathscr{P}^-$ and $\mathscr{P}^+$ in Eqs. (4.7) that the transient extinguishes after a value of the travelled distance equal to $s = 2a$, where $2a$ is the maximum length of the contact patch. After the tyre has travelled a distance equal to $s = 2a$, the solution $\boldsymbol{u}_t^-(\boldsymbol{\xi}, s)$ extends all over $\mathscr{P}$. It is, however, still time-varying, and depends on the specific expressions of the translational slip and spin parameters. If these are constant over time or, equivalently, the travelled distance, the solutions in Eqs. (4.5) further simplify to

$$\boldsymbol{u}_t^-(\boldsymbol{\xi}) = \boldsymbol{\sigma}\xi + \mathbf{A}_\varphi \xi \begin{bmatrix} x_{\mathscr{L}}(\eta) - \xi/2 \\ \eta \end{bmatrix}, \qquad (\boldsymbol{\xi}, s) \in \mathscr{P}^- \times \mathbb{R}_{\geq 0}, \tag{4.11a}$$

$$\boldsymbol{u}_t^+(\boldsymbol{\xi}, s) = \boldsymbol{\sigma}s + \mathbf{A}_\varphi s \begin{bmatrix} x_{\mathscr{L}}(\eta) - \xi + s/2 \\ \eta \end{bmatrix} + \boldsymbol{u}_{t0}(\xi - s, \eta), \quad (\boldsymbol{\xi}, s) \in \mathscr{P}^+ \times \mathbb{R}_{\geq 0}, \tag{4.11b}$$

where the stationary solution $\boldsymbol{u}_t^-(\boldsymbol{\xi})$ is also independent of s, and coincides with that already obtained in Chap. 3. In this case, in fact, the solution $\boldsymbol{u}_t^-(\boldsymbol{\xi})$ is not only stationary but also steady-state. For completeness, the solutions (4.11) may be transformed back to the original variables $\boldsymbol{x}$ yielding

$$\boldsymbol{u}_t^-(\boldsymbol{x}) = \boldsymbol{\sigma}\big(x_{\mathscr{L}}(y) - x\big) + \mathbf{A}_\varphi\big(x_{\mathscr{L}}(y) - x\big) \begin{bmatrix} \big(x_{\mathscr{L}}(y) + x\big)/2 \\ y \end{bmatrix}, \quad (\boldsymbol{x}, s) \in \mathscr{P}^- \times \mathbb{R}_{\geq 0}, \tag{4.12a}$$

$$\boldsymbol{u}_t^+(\boldsymbol{x}, s) = \boldsymbol{\sigma}s + \mathbf{A}_\varphi s \begin{bmatrix} x + s/2 \\ y \end{bmatrix} + \boldsymbol{u}_{t0}(x + s, y), \qquad (\boldsymbol{x}, s) \in \mathscr{P}^+ \times \mathbb{R}_{\geq 0}. \tag{4.12b}$$

The dynamics of the travelling edge may be studied using classic results from differential geometry (Appendix 4.B). In particular, using the original set of variables $\boldsymbol{x}$, the travelling edge may be described implicitly by

$$\gamma_\Sigma(\boldsymbol{x}, s) = x_{\mathscr{L}}(y) - x - s = 0, \qquad (\boldsymbol{x}, s) \in \mathscr{P} \times \mathbb{R}_{\geq 0}, \tag{4.13}$$

[2] Therefore, the transient brush theory may be seen as a *weak* one, in the sense that the solutions are always $C^0(\mathscr{P} \times \mathbb{R}_{\geq 0}; \mathbb{R}^2)$, but higher regularity cannot be required.

and the outward-pointing unit normal is calculated as

$$\hat{\boldsymbol{\nu}}_\Sigma(\boldsymbol{x}, s) \triangleq \pm\frac{\nabla_t\gamma_\Sigma(\boldsymbol{x}, s)}{\left\|\nabla_t\gamma_\Sigma(\boldsymbol{x}, s)\right\|} = \pm\begin{bmatrix}-1 & \dfrac{\partial x_{\mathscr{L}}(y)}{\partial y}\end{bmatrix}^{\mathrm{T}}\left(1+\left(\frac{\partial x_{\mathscr{L}}(y)}{\partial y}\right)^2\right)^{-1/2}. \tag{4.14}$$

Accordingly, a particular representation of the velocity of the travelling edge that is oriented as the unit normal is given as in Eq. (4.87):

$$\bar{\boldsymbol{v}}_\Sigma^{(\hat{\nu})}(\boldsymbol{x}, s) \triangleq -\frac{\nabla_t\gamma_\Sigma(\boldsymbol{x}, s)}{\left\|\nabla_t\gamma_\Sigma(\boldsymbol{x}, s)\right\|^2}\frac{\partial\gamma_\Sigma(\boldsymbol{x}, s)}{\partial s}. \tag{4.15}$$

For a rectangular contact patch, it may be easily verified that Eq. (4.15) gives $\bar{\boldsymbol{v}}_\Sigma^{(\hat{\nu})} = -\hat{\boldsymbol{e}}_x$, since the parametrisation of the leading edge does not depend on the lateral coordinate and is identically given by $x = x_{\mathscr{L}}(y) = a$. As a result, the bristles located at $x > a - s$ undergo simultaneously a stationary deformation, and the travelling edge is represented by a straight line parallel to the y-axis. On the other hand, for an elliptical contact patch, the travelling edge velocity $\bar{\boldsymbol{v}}_\Sigma^{(\hat{\nu})}(\boldsymbol{x})$ has two components and reads

$$\bar{\boldsymbol{v}}_\Sigma^{(\hat{\nu})}(\boldsymbol{x}) = -\begin{bmatrix}1 & \dfrac{ay}{\sqrt{b^2(b^2-y^2)}}\end{bmatrix}^{\mathrm{T}}\left(1+\frac{a^2y^2}{b^2(b^2-y^2)}\right)^{-1}. \tag{4.16}$$

From Eq. (4.16), it may be deduced that, at $y = 0$, only the component in the longitudinal direction is retained, and coincides with that of the travelled distance s. Also, in this case, the transient region $\mathscr{P}^+$ may be obtained as the intersection between the contact ellipse and another ellipse translated in the longitudinal direction by $-s$. It is also worth noticing that all the bristles located at $|y| > b\sqrt{1-(s/(2a))^2}$ are already stationary, whilst the transient extinguishes slower at $y = 0$, after exactly $\bar{s} \triangleq s/(2a) = 1$. This effect is not perceptible when using the variables $\boldsymbol{\xi}$, since the leading edge is straightened by the change of coordinates.

4.2 Effect of Limited Friction

The present section extends the transient analysis to a more general case of limited friction available inside the contact patch. In this context, closed-form solutions may be derived when considering vertical pressure distributions which are constant over time and satisfy Assumption 3.1.2. The slip and spin inputs, as well as the contact shape, are also assumed to be constant over time. Furthermore, for the case of pure translational slips, the investigation is conducted with respect to any concave pressure distribution; on the other hand, when the effect of the spin is accounted for, the pressure distribution is modelled explicitly using a parabolic trend, as in

Eq. (3.22). For all the problems at hand, well-posed solutions are considered functions at least $C^0(\mathscr{P} \times \mathbb{R}_{\geq 0}; \mathbb{R}^2)$ solving Eqs. (4.2) weakly in the adhesion zone $\mathscr{P}^{(a)}$, satisfying the BC and IC given respectively by Eqs. (4.3), (4.4) and of the form (3.19) in $\mathscr{P}^{(s)}$. In the transition from sliding to adhesion and *vice versa*, the sought solutions need to satisfy the BCs (2.19) and (2.23).

The following cases are considered: pure lateral slip, pure spin slip with non-supercritical spin, and combined lateral and spin slip conditions with subcritical spin. For all these operating conditions of the tyre, the solution may be found by simply constraining the vanishing sliding solution below the friction limit. Indeed, every solution of this type automatically satisfies the BCs (2.19) and (2.23). The analysis proposed in this book is inspired by the works authored by Kalker [14] and Romano et al. [15, 16].

4.2.1 Pure Lateral Slip

Pure lateral conditions ($\sigma_x = 0, \sigma_y \neq 0, \varphi = 0$) may be investigated in isolation considering initial conditions of the type $u_{x0}(\boldsymbol{\xi}) = 0$ for all $\boldsymbol{\xi} \in \mathscr{P}$. The latter condition allows to construct the transient solution combining the full-adhesion expressions in Eqs. (4.11) with stationary sliding solutions of the form $\boldsymbol{u}_t^{(s)}(\boldsymbol{\xi}, s) = u_y^{(s)}(\boldsymbol{\xi}, s)\hat{\boldsymbol{e}}_y$, with

$$u_y^{(s)}(\boldsymbol{\xi}) = \frac{\mu}{k} q_z(\boldsymbol{\xi}) \operatorname{sgn} \sigma_y \quad \text{if} \quad \sigma_y u_y^{(a)}\bigl(\xi_{\mathscr{S}}(\eta, s), \eta, s\bigr) \geq 0, \quad k\bigl|\sigma_y\bigr| > \mu \frac{\partial q_z(\boldsymbol{\xi})}{\partial \xi}, \tag{4.17a}$$

$$u_y^{(s)}(\boldsymbol{\xi}) = -\frac{\mu}{k} q_z(\boldsymbol{\xi}) \operatorname{sgn} \sigma_y \quad \text{if} \quad \sigma_y u_y^{(a)}\bigl(\xi_{\mathscr{S}}(\eta, s), \eta, s\bigr) < 0, \quad k\bigl|\sigma_y\bigr| < -\mu \frac{\partial q_z(\boldsymbol{\xi})}{\partial \xi}, \tag{4.17b}$$

where, as usual, $\xi = \xi_{\mathscr{S}}(\eta, s)$ parametrises explicitly a sliding edge $\mathscr{S}$ [16]. Equations (4.17a) and (4.17b) are valid when the adhesion shear stress $q_y^{(a)}(\boldsymbol{\xi}, s) = ku_y^{(a)}(\boldsymbol{\xi}, s)$ given by Eq. (4.11) exceeds the friction bound that is concordant or discordant with the sign of the slip σ_y, respectively.[3] In both cases, the sliding direction is constant and equals $\hat{\boldsymbol{s}}_t = \pm \operatorname{sgn} \sigma_y \hat{\boldsymbol{e}}_y$. In absence of slip, that is $\sigma_y = 0$, the bristle deflection is instead oriented as $\hat{\boldsymbol{s}}_t(s) = \operatorname{sgn}(u_y^{(a)}(\xi_{\mathscr{S}}(\eta, s), \eta, s))\hat{\boldsymbol{e}}_y$. The solutions constructed according to Eqs. (4.17) are continuous on the sliding edge $\mathscr{S}$, and therefore satisfy the BC (2.23).

Owing to these premises, the global solution $C^0(\mathscr{P} \times \mathbb{R}_{\geq 0}; \{0\} \times \mathbb{R})$ may be defined similarly to what was done in Chap. 3:

[3] The coordinates for which $k\bigl|\sigma_y\bigr| = \mu\partial q_z(\boldsymbol{\xi})/\partial\xi$ and $k\bigl|\sigma_y\bigr| = -\mu\partial q_z(\boldsymbol{\xi})/\partial\xi$ correspond to the points where the slope of the lateral shear stress equals that of the friction bound, and for which the micro-sliding velocity vanishes, that is $\bar{\boldsymbol{v}}_s(\boldsymbol{\xi}, s) = \boldsymbol{0}$.

$$\boldsymbol{u}_t(\boldsymbol{\xi}, s) = \begin{cases} \boldsymbol{u}_t^{(\mathrm{a})}(\boldsymbol{\xi}, s) = u_y^{(\mathrm{a})}(\boldsymbol{\xi}, s)\hat{\boldsymbol{e}}_y, & (\boldsymbol{\xi}, s) \in \mathscr{P}^{(\mathrm{a})} \times \mathbb{R}_{\geq 0}, \\ \boldsymbol{u}_t^{(\mathrm{s})}(\boldsymbol{\xi}) = u_y^{(\mathrm{s})}(\boldsymbol{\xi})\hat{\boldsymbol{e}}_y, & (\boldsymbol{\xi}, s) \in \mathscr{P}^{(\mathrm{s})} \times \mathbb{R}_{\geq 0}, \end{cases} \tag{4.18}$$

with $u_y^{(\mathrm{a})}(\boldsymbol{\xi}, s)$ reading[4]

$$u_y^{(\mathrm{a})}(\boldsymbol{\xi}, s) = \begin{cases} u_y^-(\xi) = \sigma_y \xi, & (\boldsymbol{\xi}, s) \in \mathscr{P}^- \times \mathbb{R}_{\geq 0}, \\ u_y^+(\boldsymbol{\xi}, s) = \sigma_y s + u_{y0}(\xi - s, \eta), & (\boldsymbol{\xi}, s) \in \mathscr{P}^+ \times \mathbb{R}_{\geq 0}, \end{cases} \tag{4.19}$$

and $u_y^{(\mathrm{s})}(\boldsymbol{\xi})$ as in Eqs. (4.17). Moreover, the adhesion and sliding zone in Eqs. (4.19) may be alternatively defined as

$$\mathscr{P}^{(\mathrm{a})} \triangleq \left\{ \boldsymbol{\xi} \in \mathscr{P} \,\middle|\, k\left|u_y^{(\mathrm{a})}(\boldsymbol{\xi}, s)\right| \leq \mu q_z(\boldsymbol{\xi}) \right\}, \tag{4.20a}$$

$$\mathscr{P}^{(\mathrm{s})} \triangleq \left\{ \boldsymbol{\xi} \in \mathscr{P} \,\middle|\, k\left|u_y^{(\mathrm{a})}(\boldsymbol{\xi}, s)\right| > \mu q_z(\boldsymbol{\xi}) \right\}. \tag{4.20b}$$

The latter result is a direct consequence of the condition

$$\left|q_{y0}(\boldsymbol{\xi})\right| = k\left|u_{y0}(\boldsymbol{\xi})\right| \leq \mu q_z(\boldsymbol{\xi}), \qquad \forall\, \boldsymbol{\xi} \in \mathscr{P}, \tag{4.21}$$

and the concavity of $q_z(\boldsymbol{\xi})$ in the rolling direction. This is formalised mathematically by Lemmata 4.A.1 and 4.A.3, which assert that the solution constructed according to Eqs. (4.17), (4.18) and (4.19) can never violate the conditions in the right-hand sides of Eqs. (4.17), implying that the BCs (2.19) and (2.23) are automatically satisfied on any adhesion or sliding edge.

Lemmata 4.A.1 and 4.A.3 may be interpreted graphically as shown in Fig. 4.1, where the transient solution is plotted for a value of the nondimensional travelled distance $\bar{s} = s/(2a) = 1/3$ against the nondimensional coordinate $\bar{\xi} = \xi/(2a)$, and using a parabolic pressure distribution. In Fig. 4.1, the initial conditions have been assumed to be of the form $\boldsymbol{u}_{t0}(\boldsymbol{\xi}) = u_{y0}(\boldsymbol{\xi})\hat{\boldsymbol{e}}_y$, and have discordant sign to that of the new slip value $\sigma_y > 0$. Points B and C represent the coordinates where the partial derivative of the lateral shear stress $k\sigma_y$ equals the slope of the friction bound. Lemma 4.A.1 ensures that the transient solution never exceeds the positive (negative) friction parabola before point B (C), implying that the micro-sliding velocity never vanishes [16].

Furthermore, for values of the lateral coordinate η such that $\left|\sigma_y\right| \geq \sigma^*(\eta)$, it is possible to show that the sliding solution is always concordant with the sign of the lateral slip σ_y, provided that the vertical pressure distribution is strictly concave in

[4] It should be noticed that in Eqs. (4.19) the adhesion solution $u_y^{(\mathrm{a})}(\boldsymbol{\xi}, s)$ has been extended analytically over the whole contact patch $\mathscr{P}$. This makes it possible to restate $\mathscr{P}^{(\mathrm{a})}$ and $\mathscr{P}^{(\mathrm{s})}$ as in Eqs. (4.20).

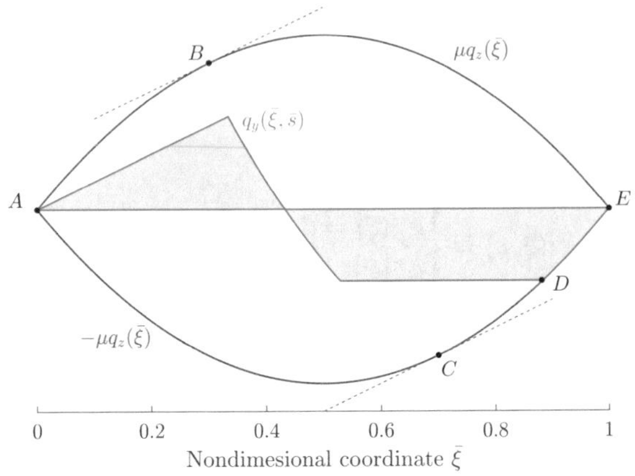

Fig. 4.1 Transient solution for $\bar{s} = 1/3$ starting from lateral initial conditions $u_{y0}(\boldsymbol{\xi})$ which have opposite sign to the new slip value

the rolling direction (Proposition 4.A.1). The same result may be generalised to the case of simply concave pressure distributions if the condition is satisfied with strictly inequality, that is $|\sigma_y| > \sigma^*(\eta)$ (Proposition 4.A.2).

Lemmata 4.A.2 and 4.A.4 also demonstrate that a bristle that slides at a certain point inside the contact patch keeps sliding until it disengages the contact with the road.

Figure 4.2 exemplifies the transient evolution of the lateral shear stress starting from initial conditions similar to those found by Cattaneo [17]. In vehicle dynamics, a similar distribution may originate from transient trends due to critical or supercritical slips, that is $|\sigma_y| \geq \sigma^{\mathrm{cr}}$ (several examples are reported in [15]). It may be clearly observed that the initial conditions are gradually faded out. Moreover, according to the classic formulation of the brush models, the transient always extinguishes completely after travelling a distance equal to the contact length.

In general, depending on the value of the lateral slip, different adhesion and sliding areas arise inside the contact patch. Some analytical results are available in [15] for zero initial conditions.

From Eq. (4.12b), the dynamics of a sliding edge $\mathscr{S}$ may be studied considering the implicit representation

$$\gamma_{\mathscr{S}}(\boldsymbol{x}, s) = k\left|u_y^+(\boldsymbol{x}, s)\right| - \mu q_z(\boldsymbol{x}) = k\left|\sigma_y s + u_{y0}(x + s, y)\right| - \mu q_z(\boldsymbol{x}) = 0, \quad (\boldsymbol{x}, s) \in \mathscr{P} \times \mathbb{R}_{\geq 0}. \tag{4.22}$$

Its motion may be described by computing the velocity of the sliding edge $\mathscr{S}$ that is oriented as the unit normal (Appendix 4.B). From Eq. (4.85), it may be calculated as

$$\bar{\boldsymbol{v}}_{\mathscr{S}}^{(\hat{\nu})}(\boldsymbol{x}, s) \triangleq -\frac{\nabla_t \gamma_{\mathscr{S}}(\boldsymbol{x}, s)}{\left\|\nabla_t \gamma_{\mathscr{S}}(\boldsymbol{x}, s)\right\|^2} \frac{\partial \gamma_{\mathscr{S}}(\boldsymbol{x}, s)}{\partial s}, \tag{4.23}$$

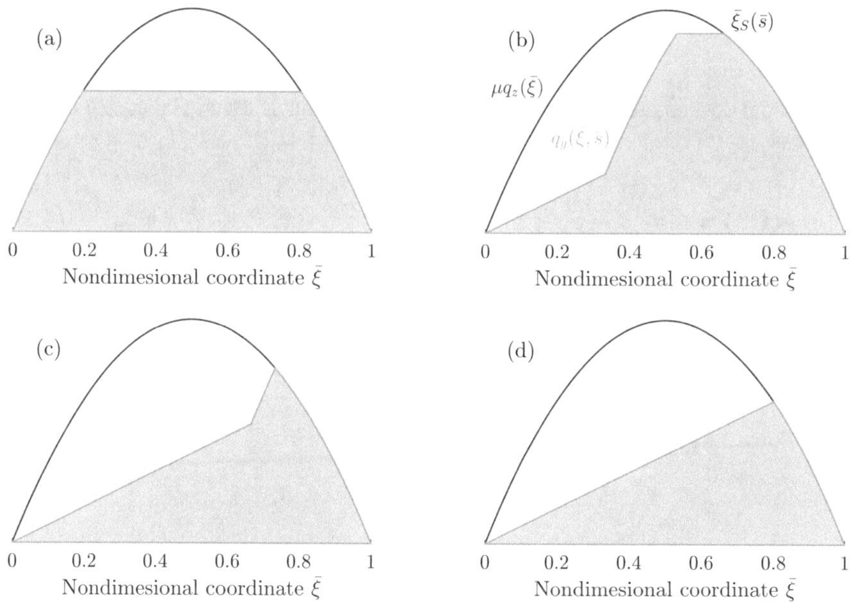

Fig. 4.2 Transient evolution of the lateral shear stress due to a lateral slip input starting from Cattaneo's initial conditions. **a** $\bar{s} = 0$; **b** $\bar{s} = 1/3$; **c** $\bar{s} = 2/3$; **d** $\bar{s} = 1$

with

$$\frac{\partial \gamma_{\mathscr{S}}(\boldsymbol{x}, s)}{\partial x} = \frac{u_y^+(\boldsymbol{x}, s)}{\left|u_y^+(\boldsymbol{x}, s)\right|} k \frac{\partial u_{y0}(x+s, y)}{\partial x} - \mu \frac{\partial q_z(\boldsymbol{x})}{\partial x}, \tag{4.24a}$$

$$\frac{\partial \gamma_{\mathscr{S}}(\boldsymbol{x}, s)}{\partial y} = \frac{u_y^+(\boldsymbol{x}, s)}{\left|u_y^+(\boldsymbol{x}, s)\right|} k \frac{\partial u_{y0}(x+s, y)}{\partial y} - \mu \frac{\partial q_z(\boldsymbol{x})}{\partial y}, \tag{4.24b}$$

$$\frac{\partial \gamma_{\mathscr{S}}(\boldsymbol{x}, s)}{\partial s} = \frac{u_y^+(\boldsymbol{x}, s)}{\left|u_y^+(\boldsymbol{x}, s)\right|} k \left(\sigma_y + \frac{\partial u_{y0}(x+s, y)}{\partial s} \right), \tag{4.24c}$$

where

$$\frac{\partial u_{y0}(x+s, y)}{\partial x} = \frac{\partial u_{y0}(x+s, y)}{\partial s}. \tag{4.25}$$

Example 4.2.1 For a rectangular contact patch with zero initial condition $u_{y0}(\xi) = 0$, the sliding edge is parametrised in implicit form as

$$\gamma_{\mathscr{S}}(\xi, s) = k\left|\sigma_y\right|s - \mu\frac{q_z^*}{a^2}\xi(2a - \xi) = 0, \qquad (\boldsymbol{\xi}, s) \in \mathscr{P} \times \mathbb{R}_{\geq 0}, \tag{4.26}$$

in the local coordinate system $\boldsymbol{\xi}$. Using the original variables $\boldsymbol{x}$, Eq. (4.26) may be recast as

$$\gamma_{\mathscr{S}}(x, s) = k\left|\sigma_y\right|s - \mu q_z^*\left[1 - \left(\frac{x}{a}\right)^2\right] = 0, \qquad (\boldsymbol{x}, s) \in \mathscr{P} \times \mathbb{R}_{\geq 0}. \tag{4.27}$$

Moreover, for $\left|\sigma_y\right| < \sigma^{\mathrm{cr}}/2$, there is only one sliding edge which may be parametrised in explicit form by

$$\xi_{\mathscr{S}}(s) = a\left(1 + \sqrt{1 - \frac{2\left|\sigma_y\right|s}{a\sigma^{\mathrm{cr}}}}\right), \qquad s \in [0, \lambda_{\mathrm{ad}}], \tag{4.28}$$

or

$$x_{\mathscr{S}}(s) = -a\sqrt{1 - \frac{2\left|\sigma_y\right|s}{a\sigma^{\mathrm{cr}}}}, \qquad s \in [0, \lambda_{\mathrm{ad}}]. \tag{4.29}$$

Calculating the velocity of the sliding edge according to Eq. (4.23) provides

$$\bar{\boldsymbol{v}}_{\mathscr{S}}^{(\hat{\nu})}(x, s) = \frac{\left|\sigma_y\right|}{\sigma^{\mathrm{cr}}x}\hat{\boldsymbol{e}}_x, \tag{4.30}$$

which, using Eq. (4.27), becomes

$$\bar{\boldsymbol{v}}_{\mathscr{S}}^{(\hat{\nu})}(s) = -\frac{\left|\sigma_y\right|}{\sigma^{\mathrm{cr}}\sqrt{1 - \frac{2\left|\sigma_y\right|s}{a\sigma^{\mathrm{cr}}}}}\hat{\boldsymbol{e}}_x = -\frac{\partial x_{\mathscr{S}}(s)}{\partial s}\hat{\boldsymbol{e}}_x = \frac{\partial \xi_{\mathscr{S}}(s)}{\partial s}\hat{\boldsymbol{e}}_x, \tag{4.31}$$

where the last two identities stem from the fact that the problem is one-dimensional in the space coordinates.

The transient analysis may be finally concluded by observing that the results stated for the case of pure lateral slip can be extended trivially to the case of combined translational slips if the initial conditions are oriented as the new slip input $\boldsymbol{\sigma}$.

4.2.2 Pure Spin Slip

The transient dynamics in the case of pure spin may be investigated under the hypothesis of a thin tyre. As in Sect. 3.4, the contact patch is assumed to be one-dimensional and described by $\mathscr{P} = \{\xi \in \mathbb{R} \mid 0 \leq \xi \leq 2a\}$, or equivalently by $\mathscr{P} = \{x \in \mathbb{R} \mid -a \leq x \leq a\}$ in the original variable x. Accordingly, the dependency on y or η is omitted to lighten the notation. The pressure distribution is supposed to be parabolic as in Eq. (3.22).

For the sake of simplicity, the investigation is also limited to the case of non-supercritical spin $|\varphi| \leq \varphi^{\mathrm{cr}}$, and zero initial longitudinal conditions $u_{x0}(\xi) = 0$. When $\varphi \neq 0$, these two conditions provide the following expressions for the lateral component of the sliding solution [16]:

$$u_y^{(\mathrm{s})}(\xi) = \frac{\mu}{k} q_z(\xi) \operatorname{sgn} \varphi \quad \text{if} \quad \varphi u_y^{(\mathrm{a})}\big(\xi_{\mathscr{S}}(s), s\big) \geq 0, \quad \big(|\varphi| - \varphi^{\mathrm{cr}}\big)(a - \xi) > 0, \tag{4.32a}$$

$$u_y^{(\mathrm{s})}(\xi) = -\frac{\mu}{k} q_z(\xi) \operatorname{sgn} \varphi \quad \text{if} \quad \varphi u_y^{(\mathrm{a})}\big(\xi_{\mathscr{S}}(s), s\big) < 0, \quad \big(|\varphi| + \varphi^{\mathrm{cr}}\big)(a - \xi) < 0, \tag{4.32b}$$

and the tangential displacement is clearly oriented as $\hat{\boldsymbol{s}}_t = \pm \operatorname{sgn} \varphi \hat{\boldsymbol{e}}_y$. The solution $\boldsymbol{u}_t(\xi, s) \in C^0(\mathscr{P} \times \mathbb{R}_{\geq 0}; \{0\} \times \mathbb{R})$ may be patched on the whole contact area as

$$u_y^{(\mathrm{a})}(\xi, s) = \begin{cases} u_y^-(\xi) = \dfrac{1}{2}\varphi\xi(2a - \xi), & (\xi, s) \in \mathscr{P}^- \times \mathbb{R}_{\geq 0}, \\ u_y^+(\xi, s) = \dfrac{1}{2}\varphi s(2a - 2\xi + s) + u_{y0}(\xi - s), & (\xi, s) \in \mathscr{P}^+ \times \mathbb{R}_{\geq 0}, \end{cases} \tag{4.33}$$

with $u_y^{(\mathrm{s})}(\xi)$ given by Eqs. (4.32), $\mathscr{P}^- = \{\xi \in \mathscr{P} \mid 0 \leq \xi < s\}$, $\mathscr{P}^+ = \{\xi \in \mathscr{P} \mid s \leq \xi \leq 2a\}$ and $\mathscr{P}^{(\mathrm{a})}$, $\mathscr{P}^{(\mathrm{s})}$ defined similarly as in Eqs. (4.20):

$$\mathscr{P}^{(\mathrm{a})} \triangleq \left\{\xi \in \mathscr{P} \;\middle|\; k\left|u_y^{(\mathrm{a})}(\xi, s)\right| \leq \mu q_z(\xi)\right\}, \tag{4.34a}$$

$$\mathscr{P}^{(\mathrm{s})} \triangleq \left\{\xi \in \mathscr{P} \;\middle|\; k\left|u_y^{(\mathrm{a})}(\xi, s)\right| > \mu q_z(\xi)\right\}. \tag{4.34b}$$

According to Eqs. (4.32) and (4.43), the analytical expression for the bristle deflection $\boldsymbol{u}_t(\xi, s)$ fulfils the two requirements on the right-hand side of (4.32), implying that nonzero values of the micro-sliding speed are only possible in the right-half of the contact patch, that is for $\xi \in (a, 2a]$. Owing to the condition $|q_{y0}(\xi)| \leq \mu q_z(\xi)$, Lemma 4.A.5 asserts that sliding never occurs for $\xi \in [0, a]$. Therefore, the lateral stress may only exceed the friction parabola in the right-half of the contact patch, where the BC (2.19) is always satisfied on a possible adhesion edge $\mathscr{A}$ by the tran-

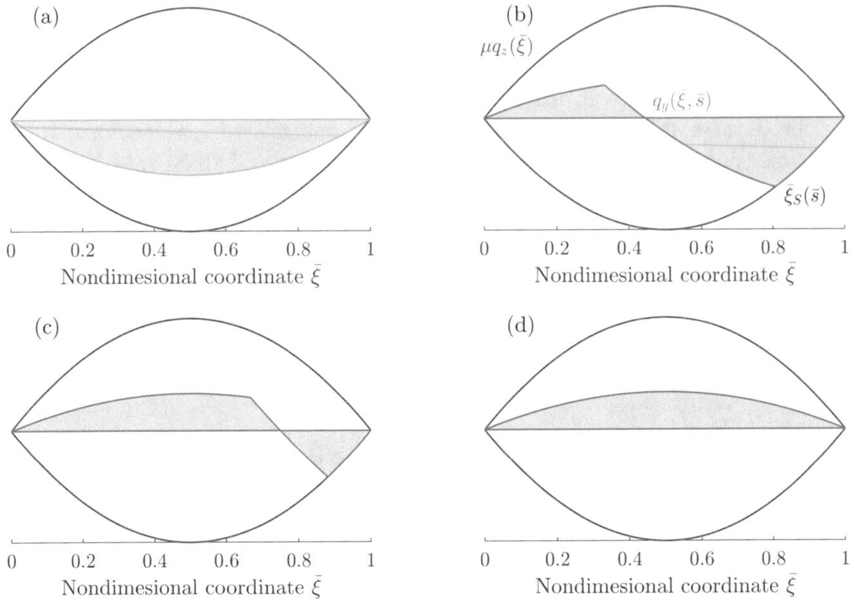

Fig. 4.3 Transient evolution of the lateral shear stress due to pure rotational slip is shown considering an initial distribution resulting from an opposite value of the spin parameter. **a** $\bar{s} = 0$; **b** $\bar{s} = 1/3$; **c** $\bar{s} = 2/3$; **d** $\bar{s} = 1$

sient solution in Eq. (4.43). This result ensures the well-posedness of the solution for any value of $|\varphi| < \varphi^{\text{cr}}$. However, in critical spin conditions, that is $|\varphi| = \varphi^{\text{cr}}$, Eq. (4.32b) states that the only possible sliding solution is that which has a discordant sign to that of the spin. In this case, Lemma 4.A.6 demonstrates that the lateral shear stress never exceeds the friction parabola which is concordant with φ.

Figure 4.3 depicts the transient trend of $q_y(\xi, s)$ due to pure spin $\varphi > 0$. The initial conditions also originate from a small spin $\varphi < 0$. The lateral shear stress evolves as the travelled distance increases but, as opposed to the pure lateral slip case, it reaches steady-state conditions exactly when $\bar{s} = 1$. Again, the analysis may be concluded by observing that a bristle that slides at some coordinate ξ and travelled distance s cannot adhere to the road again (Lemma 4.A.7).

In respect to the dynamics of the sliding edge, in this case the problem is truly one-dimensional, and the velocity of the sliding point in the original coordinates (x, s) is given by $\bar{\boldsymbol{v}}_{\mathscr{S}}^{(\hat{\boldsymbol{\nu}})}(x, s) = \bar{v}_{\mathscr{S}}^{(\hat{\nu})}(x, s)\hat{\boldsymbol{e}}_x$, with

$$\bar{v}_{\mathscr{S}}^{(\hat{\nu})}(x, s) = -\frac{\dfrac{\partial \gamma_{\mathscr{S}}(x, s)}{\partial x}}{\left|\dfrac{\partial \gamma_{\mathscr{S}}(x, s)}{\partial x}\right|^2} \frac{\partial \gamma_{\mathscr{S}}(x, s)}{\partial s}, \tag{4.35}$$

where this time the time-varying sliding edge is parametrised by

$$\gamma_{\mathscr{S}}(x,s) = k\left|u_y^+(x,s)\right| - \mu q_z(x) = k\left|\frac{1}{2}\varphi s(2x+s) + u_{y0}(x+s)\right| - \mu q_z(x) = 0,$$
$$(x,s) \in [-a,a] \times \mathbb{R}_{\geq 0}, \tag{4.36}$$

and the partial derivatives read

$$\frac{\partial \gamma_{\mathscr{S}}(x,s)}{\partial x} = \frac{u_y^+(x,s)}{\left|u_y^+(x,s)\right|}k\left(\varphi s + \frac{\partial u_{y0}(x+s)}{\partial x}\right) + k\varphi^{\mathrm{cr}}x, \tag{4.37a}$$

$$\frac{\partial \gamma_{\mathscr{S}}(x,s)}{\partial s} = \frac{u_y^+(x,s)}{\left|u_y^+(x,s)\right|}k\left(\varphi(x+s) + \frac{\partial u_{y0}(x+s)}{\partial s}\right). \tag{4.37b}$$

It may be observed that if $\gamma_{\mathscr{S}}(x,s) = 0$ is derived starting from an explicit representation yielding $x - x_{\mathscr{S}}(s) = 0$, the velocity in Eq. (4.35) becomes simply

$$\bar{v}_{\mathscr{S}}^{(\hat{\nu})}(s) = -\frac{\partial x_{\mathscr{S}}(s)}{\partial s} = \frac{\partial \xi_{\mathscr{S}}(s)}{\partial s}, \tag{4.38}$$

where $\xi = \xi_{\mathscr{S}}(s) = a - x_{\mathscr{S}}(s)$ is the explicit parametrisation of the sliding edge in the variable ξ.

Example 4.2.2 Assuming zero initial conditions, that is $u_{y0}(\xi) = 0$, the sliding edge may be parametrised in explicit form by

$$\xi_{\mathscr{S}}(s) = a - \frac{|\varphi|}{\varphi^{\mathrm{cr}}}s + \sqrt{a^2 + \frac{|\varphi|}{\varphi^{\mathrm{cr}}}\left(1 + \frac{|\varphi|}{\varphi^{\mathrm{cr}}}\right)s^2}, \qquad s \in [0, 2a], \tag{4.39}$$

using the local longitudinal variable or as

$$x_{\mathscr{S}}(s) = \frac{|\varphi|}{\varphi^{\mathrm{cr}}}s - \sqrt{a^2 + \frac{|\varphi|}{\varphi^{\mathrm{cr}}}\left(1 + \frac{|\varphi|}{\varphi^{\mathrm{cr}}}\right)s^2}, \qquad s \in [0, 2a], \tag{4.40}$$

using the original one. The scalar speed may be thus calculated by means of Eq. (4.38) and reads

$$\bar{v}_{\mathscr{S}}^{(\hat{\nu})}(s) = -\frac{\partial x_{\mathscr{S}}(s)}{\partial s} = \frac{\partial \xi_{\mathscr{S}}(s)}{\partial s} = -\frac{|\varphi|}{\varphi^{\mathrm{cr}}} + \frac{\frac{|\varphi|}{\varphi^{\mathrm{cr}}}\left(1 + \frac{|\varphi|}{\varphi^{\mathrm{cr}}}\right)s}{\sqrt{a^2 + \frac{|\varphi|}{\varphi^{\mathrm{cr}}}\left(1 + \frac{|\varphi|}{\varphi^{\mathrm{cr}}}\right)s^2}}. \tag{4.41}$$

It may be easily verified that the same result is obtained using Eq. (4.35) and substituting (4.40).

4.2.3 Lateral Slip and Spin

The transient analysis for the case combined lateral and subcritical spin slip may be worked out analogously to what already done in Sect. 3.5. The hypothesis of thin tyre should be retained, as well as that of a parabolic pressure distribution. The initial conditions are also assumed to have zero longitudinal component, that is $u_{x0}(\xi) = 0$. Owing to the previous assumptions, the lateral sliding solution may be guessed as

$$u_y^{(s)}(\xi) = \frac{\mu}{k} q_z(\xi) \operatorname{sgn} \sigma_y \quad \text{if} \quad \sigma_y u_y^{(a)}\big(\xi_{\mathscr{S}}(s), s\big) \geq 0, \quad \big|\sigma_y\big| > \Big(\varphi^{\text{cr}} - |\varphi| \operatorname{sgn}(\sigma_y \varphi)\Big)(a - \xi), \tag{4.42a}$$

$$u_y^{(s)}(\xi) = -\frac{\mu}{k} q_z(\xi) \operatorname{sgn} \sigma_y \quad \text{if} \quad \sigma_y u_y^{(a)}\big(\xi_{\mathscr{S}}(s), s\big) < 0, \quad \big|\sigma_y\big| < \Big(\varphi^{\text{cr}} + |\varphi| \operatorname{sgn}(\sigma_y \varphi)\Big)(\xi - a), \tag{4.42b}$$

and the displacement is therefore oriented as $\hat{\boldsymbol{s}}_t = \pm \operatorname{sgn} \sigma_y \hat{\boldsymbol{e}}_y$ [16]. Also in this case, the solution $\boldsymbol{u}_t(\xi, s) \in C^0(\mathscr{P} \times \mathbb{R}_{\geq 0}; \{0\} \times \mathbb{R})$ may be constructed taking inspiration from Eqs. (4.18) and (4.42), with

$$u_y^{(a)}(\xi, s) = \begin{cases} u_y^-(\xi) = \sigma_y \xi + \dfrac{1}{2}\varphi \xi(2a - \xi), & (\xi, s) \in \mathscr{P}^- \times \mathbb{R}_{\geq 0}, \\ u_y^+(\xi, s) = \sigma_y s + \dfrac{1}{2}\varphi s(2a - 2\xi + s) + u_{y0}(\xi - s), & (\xi, s) \in \mathscr{P}^+ \times \mathbb{R}_{\geq 0}, \end{cases} \tag{4.43}$$

and $\mathscr{P}^{(a)}$, $\mathscr{P}^{(s)}$ defined as in Eqs. (4.34).

Considerations about the well-posedness of the solution and the dynamics of the bristle inside the contact patch may be drawn as in Sects. 4.2.1 and 4.2.2, and are formalised by Lemmata 4.A.8, 4.A.9, 4.A.2 and 4.A.7. On the other hand, from the perspective of a physical interpretation, the formulation in terms of equivalent friction coefficient $\tilde{\mu}$ and critical slip $\tilde{\sigma}^{\text{cr}}$ provides useful insights. In particular, similarly to what was already discussed for the pure lateral case, when $\big|\sigma_y\big| > \tilde{\sigma}^{\text{cr}}$, the sliding solution is always concordant with the sign of the lateral slip (Proposition 4.A.3).

With respect to the dynamics of the sliding edge, its one-dimensional speed may be still calculated using Eq. (4.35), with $\gamma_{\mathscr{S}}(x, s) = 0$ chosen as

$$\gamma_{\mathscr{S}}(x, s) = k\big|u_y^+(x, s)\big| - \mu q_z(x) = k\left|\sigma_y s + \frac{1}{2}\varphi s(2x + s) + u_{y0}(x + s)\right| - \mu q_z(x) = 0,$$
$$(x, s) \in [-a, a] \times \mathbb{R}_{\geq 0},$$

with partial derivatives reading

$$\frac{\partial \gamma_{\mathscr{S}}(x, s)}{\partial x} = \frac{u_y^+(x, s)}{\big|u_y^+(x, s)\big|} k\left(\varphi s + \frac{\partial u_{y0}(x + s)}{\partial x}\right) + k\varphi^{\text{cr}} x, \tag{4.44a}$$

$$\frac{\partial \gamma_{\mathscr{S}}(x, s)}{\partial s} = \frac{u_y^+(x, s)}{\big|u_y^+(x, s)\big|} k\left(\sigma_y + \varphi(x + s) + \frac{\partial u_{y0}(x + s)}{\partial s}\right). \tag{4.44b}$$

Similar considerations as in Subsect. 4.2.2 yield the same results in the form of Eq. (4.38) when the sliding edge is calculated, starting from an explicit parametrisation of the type $x = x_{\mathscr{S}}(s)$.

4.3 Effect of Carcass Compliance

When the compliance of the tyre carcass is accounted for, the slip variables $\boldsymbol{\sigma}(s)$ in Eq. (4.1) may be replaced by the transient slip $\boldsymbol{\sigma}'(s)$. Assuming that adhesion conditions start at the leading edge, Eq. (4.1) may be recast as

$$\frac{\partial \boldsymbol{u}_t^{(\mathrm{a})}(\boldsymbol{\xi}, s)}{\partial s} + \frac{\partial \boldsymbol{u}_t^{(\mathrm{a})}(\boldsymbol{\xi}, s)}{\partial \xi} = \boldsymbol{\sigma}'(s) + \mathbf{A}_{\varphi}(s) \begin{bmatrix} x_{\mathscr{L}}(\eta) - \xi \\ \eta \end{bmatrix}, \quad (\boldsymbol{\xi}, s) \in \mathring{\mathscr{P}}^{(\mathrm{a})} \times \mathbb{R}_{>0}, \tag{4.45}$$

with BC and IC given again by Eqs. (4.3) and (4.4), respectively. As usual, in the sliding zone the deflection of the bristle must be determined starting from Eq. (3.19).

The problem described by the PDEs (4.45) is highly nonlinear since the transient slip itself depends on the deformation of the bristles through integration over the contact patch. However, it simplifies noticeably under the hypothesis that the sliding solution does not depend on the travelled distance, that is $\partial \boldsymbol{u}_t^{(\mathrm{s})}(\boldsymbol{\xi}, s)/\partial s = \mathbf{0}$. Owing to these premises, the transient slip $\boldsymbol{\sigma}'(s)$ may be expressed more conveniently as

$$\boldsymbol{\sigma}'(s) = \boldsymbol{\sigma}(s) - \frac{\mathrm{d}\boldsymbol{\delta}_t(s)}{\mathrm{d}s} = \boldsymbol{\sigma}(s) - \mathbf{S}'\mathbf{K}_t \iint_{\mathscr{P}^{(\mathrm{a})}(s)} \frac{\partial \boldsymbol{u}_t^{(\mathrm{a})}(\boldsymbol{\xi}, s)}{\partial s} \,\mathrm{d}\boldsymbol{\xi}, \tag{4.46}$$

where the last equality follows from the assumption $\boldsymbol{u}_t(\boldsymbol{\xi}, s) \in C^0(\mathscr{P} \times \mathbb{R}_{\geq 0})$. Combining Eq. (4.45) together with (4.46) and integrating by parts yields, after some manipulations,

$$\begin{aligned} \boldsymbol{\sigma}'(s) = {} & \left(\mathbf{I} + \mathbf{S}'\mathbf{K}_t A_{\mathscr{P}^{(\mathrm{a})}}(s)\right)^{-1} \left[\boldsymbol{\sigma}(s) - \mathbf{S}'\mathbf{K}_t \tilde{\mathbf{I}} \boldsymbol{S}_{\mathscr{P}^{(\mathrm{a})}}(s)\varphi(s)\right] \\ & + \left(\mathbf{I} + \mathbf{S}'\mathbf{K}_t A_{\mathscr{P}^{(\mathrm{a})}}(s)\right)^{-1} \mathbf{S}'\mathbf{K}_t \int_{\mathscr{S}(s)} \boldsymbol{u}_t^{(\mathrm{a})}(\boldsymbol{\xi}, s) \bar{v}_{\mathscr{S}}(\boldsymbol{\xi}, s) \cdot \hat{\boldsymbol{\nu}}_{\mathscr{S}}(\boldsymbol{\xi}, s) \,\mathrm{d}L, \end{aligned} \tag{4.47}$$

where

$$A_{\mathscr{P}^{(\mathrm{a})}}(s) \triangleq \iint_{\mathscr{P}^{(\mathrm{a})}(s)} \mathrm{d}\boldsymbol{\xi}, \tag{4.48a}$$

$$\boldsymbol{S}_{\mathscr{P}^{(\mathrm{a})}}(s) \triangleq \iint_{\mathscr{P}^{(\mathrm{a})}(s)} \begin{bmatrix} x_{\mathscr{L}}(\eta) - \xi \\ \eta \end{bmatrix} \mathrm{d}\boldsymbol{\xi}, \tag{4.48b}$$

$$\tilde{\mathbf{I}} \triangleq \begin{bmatrix} 0 & -1 \\ 1 & 0 \end{bmatrix}. \tag{4.48c}$$

In Eq. (4.47), the last term represents the line integral over the time-varying sliding edge $\mathscr{S}(s)$. The vectors $\bar{\boldsymbol{v}}_{\mathscr{S}}(\boldsymbol{\xi}, s)$ and $\hat{\boldsymbol{\nu}}_{\mathscr{S}}(\boldsymbol{\xi}, s)$ represent a nondimensional velocity of the sliding edge and its unit normal. It should be noticed that a velocity vector $\bar{\boldsymbol{v}}_{\mathscr{S}}(\boldsymbol{\xi}, s)$ is not unique (further details may be found in Appendix 4.B). However, the scalar product $\bar{\boldsymbol{v}}_{\mathscr{S}}(\boldsymbol{\xi}, s) \cdot \hat{\boldsymbol{\nu}}_{\mathscr{S}}(\boldsymbol{\xi}, s)$ may always be calculated by choosing a representation of the sliding velocity that is oriented as the unit normal, that is in the form $\bar{\boldsymbol{v}}_{\mathscr{S}}^{(\hat{\nu})}(\boldsymbol{\xi}, s)$ already encountered.

This formulation of the problem, introduced by Guiggiani [18] and called *non-linear full contact patch*, completely describes the transient dynamics of the tyre using Eq. (3.19) together with the coupled Eqs. (4.45) and (4.47). In many cases, the solution must be found numerically.

Example 4.3.1 (Rectangular contact patch with pure translational slips $\boldsymbol{\sigma} \neq \boldsymbol{0}, \varphi = 0$) For a rectangular contact patch with isotropic tread and diagonal matrix for the tyre carcass stiffness, Eq. (4.47) reduces, in scalar form, to

$$\sigma_x'(s) = \frac{C_x'\sigma_x(s) + 2bku_x\big(\xi_{\mathscr{S}}(s), s\big)}{C_x' + 2bk\xi_{\mathscr{S}}(s)}, \tag{4.49a}$$

$$\sigma_y'(s) = \frac{C_y'\sigma_y(s) + 2bku_y\big(\xi_{\mathscr{S}}(s), s\big)}{C_y' + 2bk\xi_{\mathscr{S}}(s)}, \tag{4.49b}$$

where the tyre caracass stiffnesses have been renamed $C_x' \triangleq C_{xx}'$, $C_y' \triangleq C_{yy}'$ and $C_{xy}' = C_{yx}' = 0$.

4.3.1 Pure Translational Slips

For a step slip input $\boldsymbol{\sigma}$, the problem described by Eqs. (3.19), (4.45) and (4.49) may be solved by assuming initial conditions which are oriented as the new slip value, or in isolation. Figure 4.4 shows the trend of the transient shear stresses $q_x(\xi, s)$ and $q_y(\xi, s)$ for the cases of pure longitudinal and lateral slip σ_x and $\sigma_y = 0.14$, respectively. The transient evolution of the shear stresses is similar to that obtained analytically when disregarding the effect of the tyre carcass, but steady-state conditions require longer travelled distances to be reached. In particular, it may be observed that the steady-state trend of shear stress is the same in both cases, whilst the transient ones are different for a fixed value of the nondimensional travelled distance $\bar{s}$. This is due to the fact that the bristle stiffness is the same, that is $k_x = k_y = k$, whereas the tyre carcass is stiffer in the longitudinal direction ($C_x' \gg C_y'$). The values used for Fig. 4.4 are $k = 2.67 \cdot 10^7$ N m^{-3}, and $C_x' = 6 \cdot 10^5$, $C_y' = 2.4 \cdot 10^5$ N m^{-1}. The different response to the slip input is therefore due to the anisotropy of the tyre carcass. This phenomenon is traditionally referred to as the *relaxation behaviour* of the tyre. In general, dynamic effects related to the compliance of the carcass are predominant over those of the bristles, especially in conjunction with instantaneous

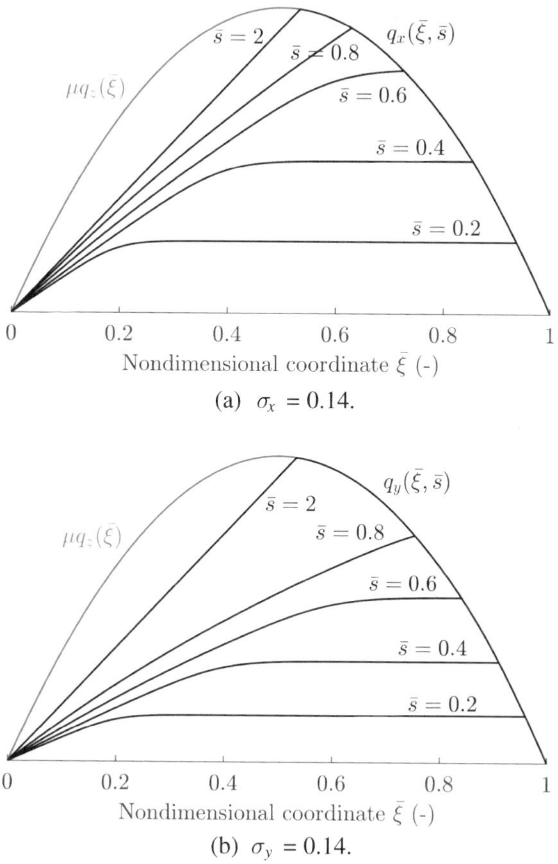

Fig. 4.4 Transient evolution of the shear stresses for pure longitudinal and lateral slip inputs

variations in the lateral slip input. However, there are several situations in which the transient deformation of the tyre tread should not be neglected [18]. These aspects will be clarified further in Chap. 6.

It should be noticed that, in solving the transient problem, the sliding solutions may be assumed to have sign equal to or opposite to the transient slips. For example, considering the pure lateral problem, the following conditions need to be satisfied for $\sigma_y'(s) \neq 0$:

$$u_y^{(\mathrm{s})}(\xi, s) = \frac{\mu}{k} q_z(\xi) \operatorname{sgn} \sigma_y'(s) \quad \text{if} \quad \sigma_y'(s) u_y^{(\mathrm{a})}\big(\xi_{\mathscr{S}}(s), s\big) \geq 0, \quad k\Big|\sigma_y'(s)\Big| > \mu \frac{\partial q_z(\xi)}{\partial \xi}, \tag{4.50a}$$

$$u_y^{(\mathrm{s})}(\xi, s) = -\frac{\mu}{k} q_z(\xi) \operatorname{sgn} \sigma_y'(s) \quad \text{if} \quad \sigma_y'(s) u_y^{(\mathrm{a})}\big(\xi_{\mathscr{S}}(s), s\big) < 0, \quad k\Big|\sigma_y'(s)\Big| < -\mu \frac{\partial q_z(\xi)}{\partial \xi}, \tag{4.50b}$$

similarly to what was already discussed for the analysis in the presence of a rigid carcass (Eqs. (4.17)). In this case, the last inequalities on the right-hand side of Eqs. (4.50) need to be checked numerically at each iteration.

4.3.2 Effect of Spin Slip

The contribution of small spin slips may be taken into account again under the hypothesis of a thin tyre. Figure 4.5 illustrates the transient trend of the lateral shear stress in combined lateral and spin slips conditions starting from undeformed initial configurations. The two subfigures refer to the cases of concordant and discordant spin slip $\varphi = \pm 0.07$ m^{-1}. In both cases, the constant lateral slip input is fixed to $\sigma_y = 0.14$. The values of the tyre parameters used to produce Fig. 4.5 are the same as earlier. The situation is again very similar to that investigated analytically in

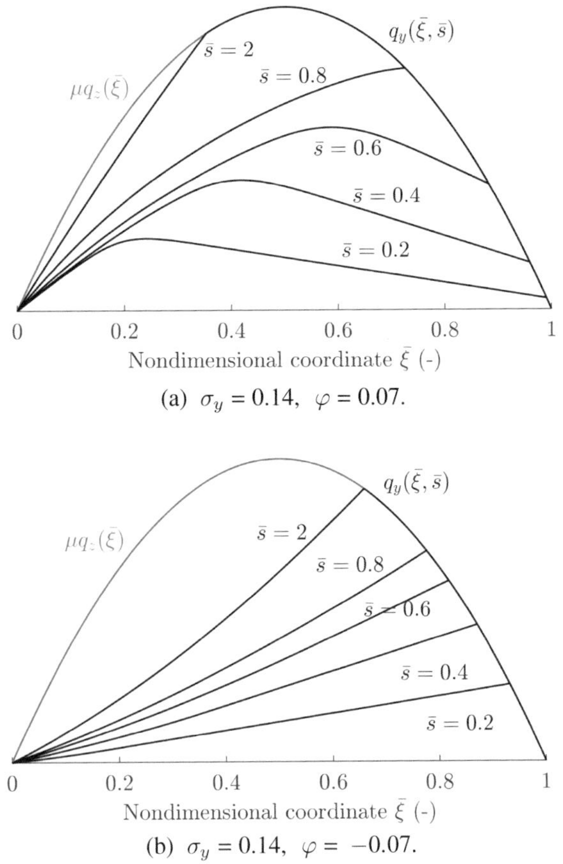

Fig. 4.5 Transient evolution of the shear stresses for combined lateral and spin slip inputs

Subsect. 4.2.3. In this case, however, the corresponding conditions of Eqs. (4.42) become

$$u_y^{(s)}(\xi) = \frac{\mu}{k} q_z(\xi) \operatorname{sgn} \sigma_y'(s) \quad \text{if} \quad \sigma_y'(s) u_y^{(a)}\big(\xi_{\mathscr{S}}(s), s\big) \geq 0, \quad \big|\sigma_y'(s)\big| > \Big(\varphi^{\text{cr}} - |\varphi| \operatorname{sgn}\big(\sigma_y'(s)\varphi\big)\Big)(a - \xi), \tag{4.51a}$$

$$u_y^{(s)}(\xi) = -\frac{\mu}{k} q_z(\xi) \operatorname{sgn} \sigma_y'(s) \quad \text{if} \quad \sigma_y'(s) u_y^{(a)}\big(\xi_{\mathscr{S}}(s), s\big) < 0, \quad \big|\sigma_y'(s)\big| < \Big(\varphi^{\text{cr}} + |\varphi| \operatorname{sgn}\big(\sigma_y'(s)\varphi\big)\Big)(\xi - a), \tag{4.51b}$$

and need to be checked iteratively.

4.A Lemmata and Propositions

4.A.1 Results for Pure Lateral Slip

The results for pure lateral slip are proved under the assumption that the pressure distribution $q_z(\boldsymbol{\xi})$ satisfies Assumption 3.1.2 and that $u_{x0}(\boldsymbol{\xi}) = 0$.

Lemma 4.A.1 *Consider pure lateral slip conditions, i.e.* $\sigma_x = 0$, $\sigma_y \neq 0$, $\varphi = 0$ *and assume that* $q_z(\boldsymbol{\xi})$ *satisfies Assumption 3.1.2 with* $q_z^{(\eta)}(\cdot)$ *strictly concave. Then, if* $\big|q_{y0}(\boldsymbol{\xi})\big| = k\big|u_{y0}(\boldsymbol{\xi})\big| \leq \mu q_z(\boldsymbol{\xi})$ *for all* $\boldsymbol{\xi} \in \mathscr{P}$*, the following implications hold for all* $(\boldsymbol{\xi}, s) \in \mathscr{P} \times \mathbb{R}_{>0}$ *such that* $\xi \in (0, \xi_{\mathscr{T}}(\eta)]$:

$$\sigma_y \geq 0 \quad \Longrightarrow \quad q_y^{(a)}(\boldsymbol{\xi}, s) < \mu q_z(\boldsymbol{\xi}), \quad k\big|\sigma_y\big| \leq \mu \frac{\partial q_z(\boldsymbol{\xi})}{\partial \xi}, \tag{4.52a}$$

$$\sigma_y \geq 0 \quad \Longrightarrow \quad q_y^{(a)}(\boldsymbol{\xi}, s) > -\mu q_z(\boldsymbol{\xi}), \quad k\big|\sigma_y\big| \geq -\mu \frac{\partial q_z(\boldsymbol{\xi})}{\partial \xi}, \tag{4.52b}$$

$$\sigma_y < 0 \quad \Longrightarrow \quad q_y^{(a)}(\boldsymbol{\xi}, s) > -\mu q_z(\boldsymbol{\xi}), \quad k\big|\sigma_y\big| \leq \mu \frac{\partial q_z(\boldsymbol{\xi})}{\partial \xi}, \tag{4.52c}$$

$$\sigma_y < 0 \quad \Longrightarrow \quad q_y^{(a)}(\boldsymbol{\xi}, s) < \mu q_z(\boldsymbol{\xi}), \quad k\big|\sigma_y\big| \geq -\mu \frac{\partial q_z(\boldsymbol{\xi})}{\partial \xi}. \tag{4.52d}$$

Proof The proofs for $\sigma_y \geq 0$ and $\sigma_y < 0$ are mirrored, and thus only the analysis for $\sigma_y \geq 0$ is conducted.

1. Consider the case $\sigma_y \geq 0$, $k\big|\sigma_y\big| \leq \mu \dfrac{\partial q_z(\boldsymbol{\xi})}{\partial \xi}$. For $\xi \in (0, s)$, Eq. (4.19) gives

$$q_y^{(a)}(\boldsymbol{\xi}, s) = q_y^{-}(\boldsymbol{\xi}) = k\big|\sigma_y\big|\xi \leq \mu \frac{\partial q_z(\boldsymbol{\xi})}{\partial \xi} \xi < \mu q_z(\boldsymbol{\xi}), \tag{4.53}$$

where the last inequality follows from Assumption 3.1.2. For $\xi \in [s, \xi_{\mathscr{T}}(\eta)]$:

$$q_y^{(a)}(\boldsymbol{\xi}, s) = q_y^{+}(\boldsymbol{\xi}, s) = k\big|\sigma_y\big| s + ku_{y0}(\xi - s, \eta) \leq \mu \frac{\partial q_z(\boldsymbol{\xi})}{\partial \xi} s + \mu q_z(\xi - s, \eta) < \mu q_z(\boldsymbol{\xi}), \tag{4.54}$$

the last inequality following from Assumption 3.1.2. Combining (4.53) and (4.54), (4.52a) is deduced.

2. Consider the case $\sigma_y \geq 0$, $k\big|\sigma_y\big| \geq -\mu \dfrac{\partial q_z(\boldsymbol{\xi})}{\partial \xi}$. For $\xi \in (0, s)$, Eq. (4.19) gives

$$q_y^{(a)}(\boldsymbol{\xi}, s) = q_y^{-}(\boldsymbol{\xi}) = k\big|\sigma_y\big| \xi \geq -\mu \frac{\partial q_z(\boldsymbol{\xi})}{\partial \xi} \xi > -\mu q_z(\boldsymbol{\xi}). \tag{4.55}$$

For $\xi \in [s, \xi_{\mathscr{T}}(\eta)]$:

$$q_y^{(a)}(\boldsymbol{\xi}, s) = q_y^{+}(\boldsymbol{\xi}, s) = k\big|\sigma_y\big| s + ku_{y0}(\xi - s, \eta) \geq -\mu \frac{\partial q_z(\boldsymbol{\xi})}{\partial \xi} s - \mu q_z(\xi - s, \eta) > -\mu q_z(\boldsymbol{\xi}), \tag{4.56}$$

the last inequality following from Assumption 3.1.2. Combining (4.55) and (4.56), (4.52b) is deduced.

Proposition 4.A.1 *Consider a vertical pressure distribution $q_z(\boldsymbol{\xi})$ satisfying Assumption 3.1.2 with $q_z^{(\eta)}(\cdot)$ strictly concave. Then, if $\big|q_{y0}(\boldsymbol{\xi})\big| = k\big|u_{y0}(\boldsymbol{\xi})\big| \leq \mu q_z(\boldsymbol{\xi})$ for all $\boldsymbol{\xi} \in \mathscr{P}$, the sliding solution is given by*

$$u_y^{(s)}(\boldsymbol{\xi}) = \frac{\mu}{k} q_z(\boldsymbol{\xi}) \operatorname{sgn} \sigma_y, \tag{4.57a}$$

for all $(\boldsymbol{\xi}, s) \in \mathscr{P} \times \mathbb{R}_{>0}$ such that $\xi \in (0, \xi_{\mathscr{T}}(\eta)]$ and $\big|\sigma_y\big| \geq \sigma^*(\eta)$.

Proof The proof may be split into two parts depending on whether $\sigma_y \geq \sigma^*(\eta)$ or $\sigma_y \leq -\sigma^*(\eta)$. Only the part for $\sigma_y \geq \sigma^*(\eta)$ will be proved; the proof for $\sigma_y \leq -\sigma^*(\eta)$ is analogous. First, it is observed that the steady-state solution $q_y^{+}(\boldsymbol{\xi}, s)$ is always concordant with the slip itself, and hence the result follows trivially. Instead, for $\xi \in [s, \xi_{\mathscr{T}}(\eta)]$:

$$\begin{aligned} q_y^{(a)}(\boldsymbol{\xi}, s) = q_y^{+}(\boldsymbol{\xi}, s) = k\big|\sigma_y\big| s + ku_{y0}(\xi - s, \eta) &\geq \mu \frac{\partial q_z(0, \eta)}{\partial \xi} s - \mu q_z(\xi - s, \eta) \\ > \mu q_z(s, \eta) - \mu q_z(\xi - s, \eta) > -\mu \frac{\partial q_z(\boldsymbol{\xi})}{\partial \xi} \xi &> -\mu q_z(\boldsymbol{\xi}). \end{aligned} \tag{4.58}$$

Lemma 4.A.2 *Consider pure lateral slip conditions, i.e. $\sigma_x = 0$, $\sigma_y \neq 0$, $\varphi = 0$, and a vertical pressure distribution $q_z(\boldsymbol{\xi})$ satisfying Assumption 3.1.2 with $q_z^{(\eta)}(\cdot)$ strictly concave. Then, the following implications hold for all $(\boldsymbol{\xi}, s) \in \mathscr{P} \times \mathbb{R}_{>0}$ such that $(\xi + \delta s, \delta s) \in (0, \xi_{\mathscr{T}}(\eta)] \times \mathbb{R}_{>0}$:*

$$\sigma_y \geq 0,\ q_y^{(a)}(\boldsymbol{\xi}, s) \geq \mu q_z(\boldsymbol{\xi}) \quad \Longrightarrow \quad q_y^{(a)}(\xi + \delta s, \eta, s + \delta s) > \mu q_z(\xi + \delta s, \eta), \tag{4.59a}$$

$$\sigma_y \geq 0,\ q_y^{(a)}(\boldsymbol{\xi}, s) \leq -\mu q_z(\boldsymbol{\xi}) \quad \Longrightarrow \quad q_y^{(a)}(\xi + \delta s, \eta, s + \delta s) < -\mu q_z(\xi + \delta s, \eta), \tag{4.59b}$$

$$\sigma_y < 0,\ q_y^{(a)}(\boldsymbol{\xi}, s) \geq \mu q_z(\boldsymbol{\xi}) \quad \Longrightarrow \quad q_y^{(a)}(\xi + \delta s, \eta, s + \delta s) > \mu q_z(\xi + \delta s, \eta), \tag{4.59c}$$

$$\sigma_y < 0,\ q_y^{(a)}(\boldsymbol{\xi}, s) \leq -\mu q_z(\boldsymbol{\xi}) \quad \Longrightarrow \quad q_y^{(a)}(\xi + \delta s, \eta, s + \delta s) < -\mu q_z(\xi + \delta s, \eta). \tag{4.59d}$$

Proof Again, the Lemma is only proved for $\sigma_y \geq 0$; the cases for $\sigma_y < 0$ are specular.

1. Consider the case $\sigma_y \geq 0, q_y^{(a)}(\boldsymbol{\xi}, s) \geq \mu q_z(\boldsymbol{\xi})$. First, it is observed that, owing to (4.52a), it must necessarily be $k|\sigma_y| > \mu \dfrac{\partial q_z(\boldsymbol{\xi})}{\partial \xi}$ to have $q_y^{(a)}(\boldsymbol{\xi}, s) \geq \mu q_z(\boldsymbol{\xi})$. Thus, recalling Assumption 3.1.2, it holds that

$$q_y^{(a)}(\xi + \delta s, \eta, s + \delta s) = k|\sigma_y|\,\delta s + q_y^{(a)}(\boldsymbol{\xi}, s) > \mu \frac{\partial q_z(\boldsymbol{\xi})}{\partial \xi}\,\delta s + \mu q_z(\boldsymbol{\xi}) > \mu q_z(\xi + \delta s, \eta). \tag{4.60}$$

2. Consider the case $\sigma_y \geq 0, q_y^{(a)}(\boldsymbol{\xi}, s) \leq -\mu q_z(\boldsymbol{\xi})$. First, it is observed that, owing to (4.52b), it must necessarily be $k|\sigma_y| < -\mu \dfrac{\partial q_z(\boldsymbol{\xi})}{\partial \xi}$ to have $q_y^{(a)}(\boldsymbol{\xi}, s) \leq -\mu q_z(\boldsymbol{\xi})$. Thus, recalling Assumption 3.1.2, it holds that

$$q_y^{(a)}(\xi + \delta s, \eta, s + \delta s) = k|\sigma_y|\,\delta s + q_y^{(a)}(\boldsymbol{\xi}, s) < -\mu \frac{\partial q_z(\boldsymbol{\xi})}{\partial \xi}\,\delta s - \mu q_z(\boldsymbol{\xi}) < -\mu q_z(\xi + \delta s, \eta). \tag{4.61}$$

Lemma 4.A.3 *Consider pure lateral slip conditions, i.e.* $\sigma_x = 0$, $\sigma_y \neq 0$, $\varphi = 0$. *Then, if* $q_z(\boldsymbol{\xi})$ *satisfies Assumption 3.1.2 and* $|q_{y0}(\boldsymbol{\xi})| = k|u_{y0}(\boldsymbol{\xi})| \leq \mu q_z(\boldsymbol{\xi})$ *for all* $\boldsymbol{\xi} \in \mathscr{P}$, *the following implications hold for all* $(\boldsymbol{\xi}, s) \in \mathscr{P} \times \mathbb{R}_{>0}$:

$$\sigma_y \geq 0 \quad \Longrightarrow \quad q_y^{(a)}(\boldsymbol{\xi}, s) \leq \mu q_z(\boldsymbol{\xi}), \quad k|\sigma_y| \leq \mu \frac{\partial q_z(\boldsymbol{\xi})}{\partial \xi}, \tag{4.62a}$$

$$\sigma_y \geq 0 \quad \Longrightarrow \quad q_y^{(a)}(\boldsymbol{\xi}, s) \geq -\mu q_z(\boldsymbol{\xi}), \quad k|\sigma_y| \geq -\mu \frac{\partial q_z(\boldsymbol{\xi})}{\partial \xi}, \tag{4.62b}$$

$$\sigma_y < 0 \quad \Longrightarrow \quad q_y^{(a)}(\boldsymbol{\xi}, s) \geq -\mu q_z(\boldsymbol{\xi}), \quad k|\sigma_y| \leq \mu \frac{\partial q_z(\boldsymbol{\xi})}{\partial \xi}, \tag{4.62c}$$

$$\sigma_y < 0 \quad \Longrightarrow \quad q_y^{(a)}(\boldsymbol{\xi}, s) \leq \mu q_z(\boldsymbol{\xi}), \quad k|\sigma_y| \geq -\mu \frac{\partial q_z(\boldsymbol{\xi})}{\partial \xi}. \tag{4.62d}$$

Proposition 4.A.2 *Consider a vertical pressure distribution* $q_z(\boldsymbol{\xi})$ *satisfying Assumption 3.1.2. Assume that for some* $\eta \in \mathscr{P}$ *it holds that* $|\sigma_y| > \sigma^*(\eta)$. *Then, if* $|q_{y0}(\boldsymbol{\xi})| = k|u_{y0}(\boldsymbol{\xi})| \leq \mu q_z(\boldsymbol{\xi})$ *for all* $\boldsymbol{\xi} \in \mathscr{P}$, *the sliding solution is given by*

$$u_y^{(s)}(\boldsymbol{\xi}) = \frac{\mu}{k} q_z(\boldsymbol{\xi}) \operatorname{sgn} \sigma_y, \tag{4.63}$$

for all $(\boldsymbol{\xi}, s) \in \mathscr{P} \times \mathbb{R}_{>0}$ *such that* $\xi \in (0, \xi_{\mathscr{T}}(\eta)]$ *and* $|\sigma_y| > \sigma^*(\eta)$.

Lemma 4.A.4 *Consider pure lateral slip conditions, i.e.* $\sigma_x = 0, \sigma_y \neq 0, \varphi = 0$, *and a vertical pressure distribution* $q_z(\boldsymbol{\xi})$ *satisfying Assumption 3.1.2. Then, the following implications hold for all* $(\boldsymbol{\xi}, s) \in \mathscr{P} \times \mathbb{R}_{>0}$ *such that* $(\xi + \delta s, \delta s) \in (0, \xi_{\mathscr{T}}(\eta)] \times \mathbb{R}_{>0}$:

$$\sigma_y \geq 0,\ q_y^{(a)}(\boldsymbol{\xi}, s) \geq \mu q_z(\boldsymbol{\xi}) \quad \Longrightarrow \quad q_y^{(a)}(\xi + \delta s, \eta, s + \delta s) \geq \mu q_z(\xi + \delta s, \eta), \tag{4.64a}$$

$$\sigma_y \geq 0,\ q_y^{(a)}(\boldsymbol{\xi}, s) \leq -\mu q_z(\boldsymbol{\xi}) \quad \Longrightarrow \quad q_y^{(a)}(\xi + \delta s, \eta, s + \delta s) \leq -\mu q_z(\xi + \delta s, \eta), \tag{4.64b}$$

$$\sigma_y < 0,\ q_y^{(a)}(\boldsymbol{\xi}, s) \geq \mu q_z(\boldsymbol{\xi}) \quad \Longrightarrow \quad q_y^{(a)}(\xi + \delta s, \eta, s + \delta s) \geq \mu q_z(\xi + \delta s, \eta), \tag{4.64c}$$

$$\sigma_y < 0,\ q_y^{(a)}(\boldsymbol{\xi}, s) \leq -\mu q_z(\boldsymbol{\xi}) \quad \Longrightarrow \quad q_y^{(a)}(\xi + \delta s, \eta, s + \delta s) \leq -\mu q_z(\xi + \delta s, \eta). \tag{4.64d}$$

Remark 1 If Assumption 3.1.2 is only satisfied with $q_z^{(\eta)} \in C^1(\mathring{\mathscr{P}}^{(\eta)}; \mathbb{R})$ for some or every fixed η, the results advocated in Lemmata 4.A.1, 4.A.3 and Propositions 4.A.1, 4.A.2 are only valid for $(\boldsymbol{\xi}, s) \in \mathscr{P} \times \mathbb{R}_{>0}$ such that $\xi \in \mathring{\mathscr{P}}^{(\eta)}$.

4.A.2 Results for Pure Spin Conditions

The following results for pure spin slip conditions are proved assuming that the pressure distribution $q_z(\xi)$ is as in Eq. (3.22) and $u_{x0}(\xi) = 0$.

Lemma 4.A.5 *Consider non-supercritical pure spin slip conditions, i.e.* $\boldsymbol{\sigma} = \mathbf{0}$, $|\varphi| \leq \varphi^{\text{cr}}$. *Then, if* $|q_{y0}(\xi)| = k|u_{y0}(\xi)| \leq \mu q_z(\xi)$ *for all* $\xi \in [0, 2a]$, *it holds that* $|q_y^{(a)}(\xi, s)| \leq \mu q_z(\xi)$ *for all* $(\xi, s) \in [0, a] \times \mathbb{R}_{>0}$. *Additionally, if* $|\varphi| < \varphi^{\text{cr}}$, *then* $|q_y^{(a)}(\xi, s)| < \mu q_z(\xi)$ *for all* $(\xi, s) \in (0, a] \times \mathbb{R}_{>0}$.

Proof The case for $\xi \in [0, s)$ is trivial and follows directly by the assumption $|\varphi| \leq \varphi^{\text{cr}}$. Instead, when $\xi \in [s, a]$

$$\begin{aligned}
\left|q_y^{(a)}(\xi, s)\right| &= k\left|\frac{1}{2}\varphi s(2a - 2\xi + s) + u_{y0}(\xi - s)\right| \leq \frac{1}{2}k|\varphi| s(2a - 2\xi + s) + k\left|u_{y0}(\xi - s)\right| \\
&= \frac{1}{2}k|\varphi| s(2a - 2\xi + s) + \left|q_{y0}(\xi - s)\right| \leq \mu\frac{q_z^*}{a^2} s(2a - 2\xi + s) + \mu q_z(\xi - s) = \mu q_z(\xi).
\end{aligned} \tag{4.65}$$

The case for $|\varphi| < \varphi^{\text{cr}}$ may be proved similarly.

Lemma 4.A.6 *Consider pure critical spin slip conditions, i.e.* $\boldsymbol{\sigma} = \mathbf{0}$, $|\varphi| = \varphi^{\mathrm{cr}}$. *Then, if* $|q_{y0}(\xi)| = k|u_{y0}(\xi)| \leq \mu q_z(\xi)$ *for all* $\xi \in [0, 2a]$, *the following implications hold for all* $(\xi, s) \in [s, 2a] \times \mathbb{R}_{>0}$:

$$\varphi = \varphi^{\mathrm{cr}} \quad \Longrightarrow \quad q_y^{(\mathrm{a})}(\xi, s) \leq \mu q_z(\xi), \tag{4.66a}$$

$$\varphi = -\varphi^{\mathrm{cr}} \quad \Longrightarrow \quad q_y^{(\mathrm{a})}(\xi, s) \geq -\mu q_z(\xi). \tag{4.66b}$$

Proof For implication (4.66a):

$$\begin{aligned} q_y^{(\mathrm{a})}(\xi, s) = q_y^{+}(\xi, s) &= \frac{1}{2}k\varphi^{\mathrm{cr}}s(2a - 2\xi + s) + ku_{y0}(\xi - s) = \mu\frac{q_z^*}{a^2}s(2a - 2\xi + s) + q_{y0}(\xi - s) \\ &\leq \mu\frac{q_z^*}{a^2}s(2a - 2\xi + s) + \mu\frac{q_z^*}{a^2}(\xi - s)(2a - \xi + s) = \mu q_z(\xi). \end{aligned} \tag{4.67}$$

For implication (4.66a):

$$\begin{aligned} q_y^{(\mathrm{a})}(\xi, s) = q_y^{+}(\xi, s) &= -\frac{1}{2}k\varphi^{\mathrm{cr}}s(2a - 2\xi + s) + ku_{y0}(\xi - s) = -\mu\frac{q_z^*}{a^2}s(2a - 2\xi + s) + q_{y0}(\xi - s) \\ &\geq -\mu\frac{q_z^*}{a^2}s(2a - 2\xi + s) - \mu\frac{q_z^*}{a^2}(\xi - s)(2a - \xi + s) = -\mu q_z(\xi). \end{aligned} \tag{4.68}$$

Lemma 4.A.7 *Consider non-supercritical pure spin slip conditions, i.e.* $\boldsymbol{\sigma} = \mathbf{0}$, $|\varphi| \leq \varphi^{\mathrm{cr}}$. *Then, if* $\left|q_y^{(\mathrm{a})}(\xi, s)\right| \geq \mu q_z(\xi)$ *for some* $\xi \in [a, 2a]$, *it holds that* $\left|q_y^{(\mathrm{a})}(\xi + \delta s, s + \delta s)\right| \geq \mu q_z(\xi + \delta s)$ *for all* $(\xi, s) \in (a, 2a] \times \mathbb{R}_{>0}$ *such that* $(\xi + \delta s, \delta s) \in (a, 2a] \times \mathbb{R}_{>0}$. *Additionally, if* $|\varphi| < \varphi^{\mathrm{cr}}$, *then* $\left|q_y^{(\mathrm{a})}(\xi + \delta s, s + \delta s)\right| > \mu q_z(\xi + \delta s)$ *for all* $(\xi, s) \in (a, 2a] \times \mathbb{R}_{>0}$ *such that* $(\xi + \delta s, \delta s) \in (a, 2a) \times \mathbb{R}_{>0}$.

Proof For $q_y^{(\mathrm{a})}(\xi, s) \geq \mu q_z(\xi)$:

$$\begin{aligned} q_y^{(\mathrm{a})}(\xi + \delta s, s + \delta s) &= \frac{1}{2}k\varphi\delta s(2a - 2\xi - \delta s) + q_y^{(\mathrm{a})}(\xi, s) \geq \mu\frac{q_z^*}{a^2}\delta s(2a - 2\xi - \delta s) + \mu q_z(\xi) \\ &= \mu\frac{q_z^*}{a^2}(\xi + \delta s)(2a - \xi - \delta s) = \mu q_z(\xi + \delta s), \end{aligned} \tag{4.69}$$

since $\xi > a \implies \xi > a - \delta s/2$ for all $\delta s > 0$. With the same reasoning, it is easy to show that, for $q_y^{(\mathrm{a})}(\xi, s) \leq -\mu q_z(\xi)$, it is

$$\begin{aligned} q_y^{(\mathrm{a})}(\xi + \delta s, s + \delta s) &= \frac{1}{2}k\varphi\delta s(2a - 2\xi - \delta s) + q_y^{(\mathrm{a})}(\xi, s) \leq -\mu\frac{q_z^*}{a^2}\delta s(2a - 2\xi - \delta s) - \mu q_z(\xi) \\ &= -\mu\frac{q_z^*}{a^2}(\xi + \delta s)(2a - \xi - \delta s) = -\mu q_z(\xi + \delta s). \end{aligned} \tag{4.70}$$

Combining (4.69) and (4.70) the first result follows automatically. The case for $|\varphi| < \varphi^{\mathrm{cr}}$ may be proved similarly.

4.A.3 Results for Combined Lateral Slip and Subcritical Spin

The results for combined lateral and spin slip conditions are proved assuming that the pressure distribution $q_z(\xi)$ is as in Eq. (3.22) and $u_{x0}(\xi) = 0$.

Lemma 4.A.8 *Consider combined lateral and spin slips conditions with subcritical spin, i.e.* $\sigma_x = 0$, $\sigma_y \neq 0$, $|\varphi| < \varphi^{\text{cr}}$. *Then, if* $|q_{y0}(\xi)| = k|u_{y0}(\xi)| \leq \mu q_z(\xi)$ *for all* $\xi \in [0, 2a]$, *the following implications hold for every* $(\xi, s) \in (0, 2a] \times \mathbb{R}_{>0}$:

$$\sigma_y > 0 \quad \Longrightarrow \quad q_y^{(\text{a})}(\xi, s) < \mu q_z(\xi), \quad |\sigma_y| \leq \Big(\varphi^{\text{cr}} - |\varphi| \operatorname{sgn}(\sigma_y \varphi)\Big)(a - \xi), \tag{4.71a}$$

$$\sigma_y > 0 \quad \Longrightarrow \quad q_y^{(\text{a})}(\xi, s) > -\mu q_z(\xi), \quad |\sigma_y| \geq \Big(\varphi^{\text{cr}} + |\varphi| \operatorname{sgn}(\sigma_y \varphi)\Big)(\xi - a), \tag{4.71b}$$

$$\sigma_y < 0 \quad \Longrightarrow \quad q_y^{(\text{a})}(\xi, s) > -\mu q_z(\xi), \quad |\sigma_y| \leq \Big(\varphi^{\text{cr}} - |\varphi| \operatorname{sgn}(\sigma_y \varphi)\Big)(a - \xi), \tag{4.71c}$$

$$\sigma_y < 0 \quad \Longrightarrow \quad q_y^{(\text{a})}(\xi, s) < \mu q_z(\xi), \quad |\sigma_y| \geq \Big(\varphi^{\text{cr}} + |\varphi| \operatorname{sgn}(\sigma_y \varphi)\Big)(\xi - a). \tag{4.71d}$$

Proof Only the cases for $\sigma_y > 0$, i.e. implications (4.71a) and (4.71b), will be proved; the cases for $\sigma_y < 0$ may be proved similarly.

1. To prove (4.71a), coordinates $\xi \in (0, 2a]$ such that $|\sigma_y| \leq (\varphi^{\text{cr}} - |\varphi| \operatorname{sgn}(\sigma_y \varphi))$ $(a - \xi)$ need to be considered. For $\xi \in (0, s)$:

$$\begin{aligned} q_y^{(\text{a})}(\xi, s) = q_y^-(\xi) &= k|\sigma_y|\,\xi + \frac{1}{2}k|\varphi|\,\xi(2a - \xi) \operatorname{sgn}(\sigma_y \varphi) \\ &\leq 2\mu \frac{q_z^*}{a^2}\xi(a - \xi) + \frac{1}{2}k|\varphi|\,\xi^2 \operatorname{sgn}(\sigma_y \varphi) \\ &< 2\mu \frac{q_z^*}{a^2}\xi(a - \xi) + \mu \frac{q_z^*}{a^2}\xi^2 = \mu \frac{q_z^*}{a^2}\xi(2a - \xi) = \mu q_z(\xi). \end{aligned} \tag{4.72}$$

   For $\xi \in [s, 2a]$:

$$\begin{aligned} q_y^{(\text{a})}(\xi, s) = q_y^+(\xi, s) &= k|\sigma_y|\,s + \frac{1}{2}k|\varphi|\,s(2a - 2\xi + s) \operatorname{sgn}(\sigma_y \varphi) + k u_{y0}(\xi - s) \\ &\leq 2\mu \frac{q_z^*}{a^2}s(a - \xi) + \frac{1}{2}k|\varphi|\,s^2 \operatorname{sgn}(\sigma_y \varphi) + \mu \frac{q_z^*}{a^2}(\xi - s)(2a - \xi + s) \\ &< 2\mu \frac{q_z^*}{a^2}s(a - \xi) + \mu \frac{q_z^*}{a^2}s^2 + \mu \frac{q_z^*}{a^2}(\xi - s)(2a - \xi + s) \\ &= \mu \frac{q_z^*}{a^2}\xi(2a - \xi) = \mu q_z(\xi). \end{aligned} \tag{4.73}$$

   Combining (4.72) and (4.73), (4.71a) is deduced.

2. To prove (4.71b), coordinates $\xi \in (0, 2a]$ such that $|\sigma_y| \geq (\varphi^{cr} + |\varphi| \operatorname{sgn}(\sigma_y\varphi))(\xi - a)$ need to be considered. For $\xi \in (0, s)$:

$$\begin{aligned} q_y^{(a)}(\xi, s) = q_y^-(\xi) &= k|\sigma_y|\xi + \frac{1}{2}k|\varphi|\xi(2a - \xi)\operatorname{sgn}(\sigma_y\varphi) \\ &\geq -2\mu\frac{q_z^*}{a^2}\xi(a - \xi) + \frac{1}{2}k|\varphi|\xi^2\operatorname{sgn}(\sigma_y\varphi) \\ &> -2\mu\frac{q_z^*}{a^2}\xi(a - \xi) - \mu\frac{q_z^*}{a^2}\xi^2 = -\mu\frac{q_z^*}{a^2}\xi(2a - \xi) = -\mu q_z(\xi). \end{aligned} \tag{4.74}$$

For $\xi \in [s, 2a]$:

$$\begin{aligned} q_y^{(a)}(\xi, s) = q_y^+(\xi, s) &= k|\sigma_y|s + \frac{1}{2}k|\varphi|s(2a - 2\xi + s)\operatorname{sgn}(\sigma_y\varphi) + ku_{y0}(\xi - s) \\ &\geq -2\mu\frac{q_z^*}{a^2}s(a - \xi) + \frac{1}{2}k|\varphi|s^2\operatorname{sgn}(\sigma_y\varphi) - \mu\frac{q_z^*}{a^2}(\xi - s)(2a - \xi + s) \\ &> -2\mu\frac{q_z^*}{a^2}s(a - \xi) - \mu\frac{q_z^*}{a^2}s^2 - \mu\frac{q_z^*}{a^2}(\xi - s)(2a - \xi + s) \\ &= -\mu\frac{q_z^*}{a^2}\xi(2a - \xi) = -\mu q_z(\xi). \end{aligned} \tag{4.75}$$

Combining (4.74) and (4.75), (4.71b) is deduced. Finally, the proofs for $\sigma_y < 0$ may be easily obtained by noticing that it is always possible to write $q_y^-(\xi) = -k|\sigma_y|\xi - \frac{1}{2}k|\varphi|\xi(2a - \xi)\operatorname{sgn}(\sigma_y\varphi)$ and $q_y^+(\xi, s) = -k|\sigma_y|s - \frac{1}{2}k|\varphi|s(2a - 2\xi + s)\operatorname{sgn}(\sigma_y\varphi) + ku_{y0}(\xi - s)$ when $\sigma_y < 0$.

Proposition 4.A.3 *Consider combined lateral and spin slips conditions with* $|\varphi| < \varphi^{cr}$ *and* $|\sigma_y| \geq \tilde{\sigma}^{cr}$. *Then, if* $|q_{y0}(\xi)| = k|u_{y0}(\xi)| \leq \mu q_z(\xi)$ *for all* $\xi \in [0, 2a]$, *the lateral component of the sliding solution is given by*

$$u_y^{(s)}(\xi) = \frac{\mu}{k}q_z(\xi)\operatorname{sgn}\sigma_y, \tag{4.76}$$

for all $(\xi, s) \in [0, 2a] \times \mathbb{R}_{>0}$ *such that* $\xi \in (0, 2a]$.

Proof Only the case for $\sigma_y > 0$ is proved; the proof for $\sigma_y < 0$ is analogous. First, it is observed that the steady-state shear stress is always concordant with the slip itself, and thus the result follows trivially for $\xi \in (0, s)$. On the other hand, for $\xi \in [s, 2a]$:

$$\begin{aligned}
q_y^{(a)}(\xi, s) = q_y^{+}(\xi, s) &= k|\sigma_y| s + \frac{1}{2}k|\varphi| s(2a - 2\xi + s)\,\mathrm{sgn}(\sigma_y\varphi) + ku_{y0}(\xi - s) \\
&\geq k\tilde{\sigma}^{\mathrm{cr}} s + \frac{1}{2}k|\varphi| s(2a - 2\xi + s)\,\mathrm{sgn}(\sigma_y\varphi) + ku_{y0}(\xi - s) \\
&= k\sigma^{\mathrm{cr}} s - \frac{1}{2}k|\varphi| s(2\xi - s)\,\mathrm{sgn}(\sigma_y\varphi) + q_{y0}(\xi - s) \\
&> k\sigma^{\mathrm{cr}} s - \frac{1}{2}k|\varphi|^{\mathrm{cr}} s(2\xi - s) + q_{y0}(\xi - s) \\
&= 2\mu\frac{q_z^*}{a} s - \mu\frac{q_z^*}{a^2} s(2\xi - s) - \mu\frac{q_z^*}{a^2}(\xi - s)(2a - \xi + s) = \mu q_z(\xi).
\end{aligned} \tag{4.77}$$

Since $q_y^{(a)}(\xi, s)$ is always greater than the friction parabola which has the opposite sign to the slip, the lateral component of the sliding solution must be concordant with the lateral slip itself.

Lemma 4.A.9 *Consider combined lateral and spin slips conditions with subcritical spin, i.e.* $\sigma_x = 0, \sigma_y \neq 0, |\varphi| < \varphi^{\mathrm{cr}}$. *Then, for every* $(\xi, s) \in (0, 2a] \times \mathbb{R}_{>0}$ *such that* $(\xi + \delta s, \delta s) \in (0, 2a] \times \mathbb{R}_{>0}$, *the following implications hold:*

$$\sigma_y > 0,\ q_y^{(a)}(\xi, s) \geq \mu q_z(\xi) \quad \Longrightarrow \quad q_y^{(a)}(\xi + \delta s, s + \delta s) > \mu q_z(\xi + \delta s), \tag{4.78a}$$

$$\sigma_y > 0,\ q_y^{(a)}(\xi, s) \leq -\mu q_z(\xi) \quad \Longrightarrow \quad q_y^{(a)}(\xi + \delta s, s + \delta s) < -\mu q_z(\xi + \delta s), \tag{4.78b}$$

$$\sigma_y < 0,\ q_y^{(a)}(\xi, s) \geq \mu q_z(\xi) \quad \Longrightarrow \quad q_y^{(a)}(\xi + \delta s, s + \delta s) > \mu q_z(\xi + \delta s), \tag{4.78c}$$

$$\sigma_y < 0,\ q_y^{(a)}(\xi, s) \leq -\mu q_z(\xi) \quad \Longrightarrow \quad q_y^{(a)}(\xi + \delta s, s + \delta s) < -\mu q_z(\xi + \delta s). \tag{4.78d}$$

Proof Again, only the cases for $\sigma_y > 0$, i.e. implications (4.78a) and (4.78b), will be proved; the cases for $\sigma_y < 0$ may be proved similarly.

1. To prove (4.78a), it may be observed that, owing to (4.71a), it must necessarily be $|\sigma_y| > (\varphi^{\mathrm{cr}} - |\varphi|\,\mathrm{sgn}(\sigma_y\varphi))(a - \xi)$ to have $q_y^{(a)}(\xi, s) \geq \mu q_z(\xi)$. Therefore,

$$\begin{aligned}
q_y^{(a)}(\xi + \delta s, s + \delta s) &= k|\sigma_y|\,\delta s + \frac{1}{2}k|\varphi|\,\delta s(2a - 2\xi - \delta s)\,\mathrm{sgn}(\sigma_y\varphi) + q_y^{(a)}(\xi, s) \\
&> 2\mu\frac{q_z^*}{a^2}\delta s(a - \xi) - \frac{1}{2}k|\varphi|\,\delta s^2\,\mathrm{sgn}(\sigma_y\varphi) + \mu\frac{q_z^*}{a^2}\xi(2a - \xi) \\
&> 2\mu\frac{q_z^*}{a^2}\delta s(a - \xi) - \mu\frac{q_z^*}{a^2}\delta s^2 + \mu\frac{q_z^*}{a^2}\xi(2a - \xi) \\
&= \mu\frac{q_z^*}{a^2}(\xi + \delta s)(2a - \xi - \delta s) = \mu q_z(\xi + \delta s).
\end{aligned} \tag{4.79}$$

2. To prove (4.78b), it may be observed that, owing to (4.71b), it must necessarily be $|\sigma_y| < (\varphi^{\mathrm{cr}} + |\varphi|\,\mathrm{sgn}(\sigma_y\varphi))(\xi - a)$ to have $q_y^{(a)}(\xi, s) \leq -\mu q_z(\xi)$. Therefore,

$$
\begin{aligned}
q_y^{(a)}(\xi + \delta s, s + \delta s) &= k\big|\sigma_y\big|\,\delta s + \frac{1}{2}k\big|\varphi\big|\,\delta s(2a - 2\xi - \delta s)\,\mathrm{sgn}\big(\sigma_y\varphi\big) + q_y^{(a)}(\xi, s) \\
&< -2\mu\frac{q_z^*}{a^2}\delta s(a - \xi) - \frac{1}{2}k\big|\varphi\big|\,\delta s^2\,\mathrm{sgn}\big(\sigma_y\varphi\big) - \mu\frac{q_z^*}{a^2}\xi(2a - \xi) \\
&< -2\mu\frac{q_z^*}{a^2}\delta s(a - \xi) + \mu\frac{q_z^*}{a^2}\delta s^2 - \mu\frac{q_z^*}{a^2}\xi(2a - \xi) \\
&= -\mu\frac{q_z^*}{a^2}(\xi + \delta s)(2a - \xi - \delta s) = -\mu q_z(\xi + \delta s).
\end{aligned}
\tag{4.80}
$$

The cases for $\sigma_y < 0$ may be proved analogously by noticing that it is always possible to write $q_y^{(a)}(\xi + \delta s, s + \delta s) = -k\big|\sigma_y\big|\,\delta s - \frac{1}{2}k\big|\varphi\big|\,\delta s(2a - 2\xi - \delta s)$ $\mathrm{sgn}(\sigma_y\varphi) + q_y^{(a)}(\xi, s)$ when $\sigma_y < 0$.

4.B Sliding and Travelling Edge Dynamics

To derive an expression for the velocity of a sliding edge, some basic notions from differential geometry are required. To start, it should be noted that, for a generic $\mathscr{S}$, the product $\bar{\boldsymbol{v}}_{\mathscr{S}}(\boldsymbol{x}, s) \cdot \hat{\boldsymbol{\nu}}_{\mathscr{S}}(\boldsymbol{x}, s)$ represents the normal component of the velocity of the sliding edge. This may be represented in implicit form as in Eq. (2.21). The outward-pointing unit normal to $\mathscr{S}$ is thus given by

$$
\hat{\boldsymbol{\nu}}_{\mathscr{S}}(\boldsymbol{x}, s) = \pm\frac{\nabla_t\gamma_{\mathscr{S}}(\boldsymbol{x}, s)}{\big\|\nabla_t\gamma_{\mathscr{S}}(\boldsymbol{x}, s)\big\|}. \tag{4.81}
$$

Furthermore, differentiating (4.13) with respect to the travelled distance following a point on the sliding edge yields [19, 20]

$$
\frac{\partial\gamma_{\mathscr{S}}(\boldsymbol{x}, s)}{\partial s} + \bar{\boldsymbol{v}}_{\mathscr{S}}^{(\rho)}(\boldsymbol{\rho}, s) \cdot \nabla_t\gamma_{\mathscr{S}}(\boldsymbol{x}, s) = 0, \tag{4.82}
$$

for some representation of the velocity $\bar{\boldsymbol{v}}_{\mathscr{S}}^{(\rho)}(\boldsymbol{\rho}, s)$. Therefore,

$$
\bar{\boldsymbol{v}}_{\mathscr{S}}^{(\rho)}(\boldsymbol{\rho}, s) \cdot \hat{\boldsymbol{\nu}}_{\mathscr{S}}(\boldsymbol{x}, s) = \bar{\boldsymbol{v}}_{\mathscr{S}}(\boldsymbol{x}, s) \cdot \hat{\boldsymbol{\nu}}_{\mathscr{S}}(\boldsymbol{x}, s) = \mp\frac{1}{\big\|\nabla_t\gamma_{\mathscr{S}}(\boldsymbol{x}, s)\big\|}\frac{\partial\gamma_{\mathscr{S}}(\boldsymbol{x}, s)}{\partial s}. \tag{4.83}
$$

In particular, the partial derivative $\partial\gamma_{\mathscr{S}}(\boldsymbol{x}, s)/\partial s$ reads

$$
\begin{aligned}
\frac{\partial\gamma_{\mathscr{S}}(\boldsymbol{x}, s)}{\partial s} = {} & \frac{k_{xx}u_x^{(a)}(\boldsymbol{x}, s) + k_{xy}u_y^{(a)}(\boldsymbol{x}, s)}{\big\|\mathbf{K}_t\boldsymbol{u}_t^{(a)}(\boldsymbol{x}, s)\big\|}\left(k_{xx}\frac{\partial u_x^{(a)}(\boldsymbol{x}, s)}{\partial s} + k_{xy}\frac{\partial u_y^{(a)}(\boldsymbol{x}, s)}{\partial s}\right) \\
& + \frac{k_{yx}u_x^{(a)}(\boldsymbol{x}, s) + k_{yy}u_y^{(a)}(\boldsymbol{x}, s)}{\big\|\mathbf{K}_t\boldsymbol{u}_t^{(a)}(\boldsymbol{x}, s)\big\|}\left(k_{yx}\frac{\partial u_x^{(a)}(\boldsymbol{x}, s)}{\partial s} + k_{yy}\frac{\partial u_y^{(a)}(\boldsymbol{x}, s)}{\partial s}\right) - \mu\frac{\partial q_z(\boldsymbol{x}, s)}{\partial s}.
\end{aligned}
\tag{4.84}
$$

A particular representation of the velocity of a sliding edge that is oriented as the unit normal is thus given by

$$\bar{\boldsymbol{v}}_{\mathscr{S}}^{(\hat{\boldsymbol{\nu}})}(\boldsymbol{x}, s) \triangleq -\frac{\nabla_t \gamma_{\mathscr{S}}(\boldsymbol{x}, s)}{\left\|\nabla_t \gamma_{\mathscr{S}}(\boldsymbol{x}, s)\right\|^2} \frac{\partial \gamma_{\mathscr{S}}(\boldsymbol{x}, s)}{\partial s}. \tag{4.85}$$

Analogously, for a travelling edge, similar equations yield the following expression for the unit normal:

$$\hat{\boldsymbol{\nu}}_{\Sigma}(\boldsymbol{x}, s) = \pm \frac{\nabla_t \gamma_{\Sigma}(\boldsymbol{x}, s)}{\left\|\nabla_t \gamma_{\Sigma}(\boldsymbol{x}, s)\right\|}, \tag{4.86}$$

whilst a nondimensional velocity vector which is oriented as the unit normal may be computed as

$$\bar{\boldsymbol{v}}_{\Sigma}^{(\hat{\boldsymbol{\nu}})}(\boldsymbol{x}, s) \triangleq -\frac{\nabla_t \gamma_{\Sigma}(\boldsymbol{x}, s)}{\left\|\nabla_t \gamma_{\Sigma}(\boldsymbol{x}, s)\right\|^2} \frac{\partial \gamma_{\Sigma}(\boldsymbol{x}, s)}{\partial s}. \tag{4.87}$$

References

1. Schlippe B von, Dietrich R (1941) Das flattern eined bepneuten rades. Bericht 140 der Lilienthal Gesellschaft. NACA TM 1365
2. Force and moment response of pneumatic tires to lateral motion inputs. Trans ASME, J Eng Ind 88B (1966)
3. Pauwelussen JP (2004) The local contact between Tyre and road under steady state combined slip conditions. Veh Syst Dyn 41(1):1–26. http://doi.org/10.1076/vesd.41.1.1.23406
4. Takács D, Orosz G, Stèpán G (2009) Delay effects in shimmy dynamics of wheels with stretched string-like tyres. Eur J Mech A Solids 28(3):516–525
5. Takács D, Stèpán G (2012) Micro-shimmy of towed structures in experimentally uncharted unstable parameter domain. Veh Syst Dyn 50(11):1613–1630
6. Takács D, Stèpán G, Hogan SJ (2008) Isolated large amplitude periodic motions of towed rigid wheels. Nonlinear Dyn 52:27–34. https://doi.org/10.1007/s11071-007-9253-y
7. Takács D, Stèpán G (2009) Experiments on quasiperiodic wheel shimmy. ASME J Comput Nonlinear Dynam 4(3):031007. https://doi.org/10.1115/1.3124786
8. Takács D, Stèpán G (2013) Contact patch memory of tyres leading to lateral vibrations of four-wheeled vehicles. Phil Trans R Soc A:37120120427. http://doi.org/10.1098/rsta.2012.0427
9. Besselink IJM (2000) Shimmy of aircraft main landing gears [doctoral thesis]. Delft
10. Ran S (2016) Tyre models for shimmy analysis: from linear to nonlinear [doctoral thesis]. Eindhoven
11. Pacejka HB (1966) The wheel shimmy phenomenon: a theoretical and experimental investigation with particular reference to the non-linear problem [doctoral thesis]. Delft
12. van Zanten A, Ruf WD, Lutz A (1989) Measurement and simulation of transient tire forces. SAE Technical Paper 890640. https://doi.org/10.4271/890640
13. Mavros G, Rahnejat H, King PD (2004) Transient analysis of tyre friction generation using a brush model with interconnected viscoelastic bristles. In: Wolfson school of mechanical and manufacturing engineering, Loughborough University, Loughborough, UK. https://doi.org/10.1243/146441905X9908

14. Kalker JJ (1997) Survey of wheel-rail rolling contact theory. Vehicle Syst Dyn 84:317–358. https://doi.org/10.1080/00423117908968610
15. Romano L, Bruzelius F, Jacobson B (2020) Unsteady-state brush theory. Veh Syst Dyn:1-29. https://doi.org/10.1080/00423114.2020.1774625
16. Romano L, Timpone F, Bruzelius F, Jacobson B. Analytical results in transient brush tyre models: theory for large camber angles and classic solutions with limited friction. Meccanica. https://doi.org/10.1007/s11012-021-01422-3
17. Cattaneo C (2008) Sul contatto di due corpi elastici: distribuzione locale degli sforzi. Rendiconti dell'Accademia Naturale dei Lincei. Serie 6, 227, 342–348, 434–436, 474–478
18. Guiggiani M (2018) The science of vehicle dynamics, 2nd edn. Springer International, Cham(Switzerland)
19. Truesdell C, Toupin RA (1960) The classical field theories. In: Flügge S (ed) Handbuch der Physik, vol 3/1. Berlin, Springer, p 226
20. Truesdell C, Rajagopal KR (2000) An introduction to the mechanics of fluids. Birkhäuser Boston . https://doi.org/10.1007/978-0-8176-4846-6

Chapter 5
Theory for Large Spin Slips

Abstract Large camber angles excite a two-dimensional velocity field inside the contact patch. This phenomenon causes the camber and turn spin slips to induce different effects on both the transient and steady-state tyre behaviours. As a consequence, the total spin slip is not sufficient alone to describe the tyre characteristics, and the two contributions must be regarded separately. The present analysis addresses the general problem under the assumption of a rigid carcass. First, it is shown how to solve the governing PDEs of the brush model in vanishing sliding conditions. Closed-form solutions for the bristle deflection are then derived explicitly for rectangular and elliptical contact geometries. The analysis is extended qualitatively to the situation of limited friction available inside the contact patch.

When the camber angles are sufficiently large to excite a two-dimensional velocity field in the contact patch, the camber and turn spin produce different effects and need to be treated separately [1, 2]. This aspect was already anticipated in Chap. 1, where a distinction was explicitly made between the contributions of the spin parameters φ_γ and φ_ψ. In this case, the approximation $\bar{\boldsymbol{v}}_t(\boldsymbol{x}, s) \approx \bar{\boldsymbol{v}}_t = -\hat{\boldsymbol{e}}_x$ does not hold anymore, and the dynamics of the tyre is governed by the coupled PDEs (2.10). Owing to these premises, the problem becomes substantially more complex even in its steady-state formulation, but a rigorous analysis is still possible.

Therefore, this chapter addresses the most general theory which considers the exact two-dimensional velocity field. Specifically, in Sect. 5.1 the problem is first solved under the assumption of vanishing sliding. Both time-varying slip inputs and contact patch are considered. However, throughout the entire analysis, the dynamics of the tyre carcass is systematically neglected. Explicit solutions are then provided for the case of constant slip inputs and fixed contact patch, in particular with respect to rectangular and elliptical geometries. Finally, in Sect. 5.2, the investigation is extended qualitatively to the case of limited friction. Only steady-state conditions are considered.

L. Romano, *Advanced Brush Tyre Modelling*,
SpringerBriefs in Applied Sciences and Technology,
https://doi.org/10.1007/978-3-030-98435-9_5

5.1 Vanishing Sliding

The analysis for the exact theory is conducted using the original space variables $\boldsymbol{x}$. This choice is motivated by the fact that the trajectories of the bristles in the contact patch are not known *a priori*, since $\bar{\boldsymbol{v}}_t(\boldsymbol{x}, s)$ may vary not only in space but also over s[1]. Assuming vanishing sliding conditions ($\mathscr{P}^{(\mathrm{a})} \equiv \mathscr{P}$), Eq. (2.14) simplifies to

$$\frac{\partial \boldsymbol{u}_t(\boldsymbol{x}, s)}{\partial s} + \big(\bar{\boldsymbol{v}}_t(\boldsymbol{x}, s) \cdot \nabla_t\big)\boldsymbol{u}_t(\boldsymbol{x}, s) = \boldsymbol{\sigma}(s) + \mathbf{A}_{\varphi}(s)\big(\boldsymbol{x} + \chi_{\psi}(s)\boldsymbol{u}_t(\boldsymbol{x}, s)\big), \quad (\boldsymbol{x}, s) \in \mathring{\mathscr{P}} \times \mathbb{R}_{>0}, \tag{5.1}$$

with $\bar{\boldsymbol{v}}_t(\boldsymbol{x}, s)$ as in Eq. (2.12) and BC and IC reading

$$\text{BC:} \quad \boldsymbol{u}_t(\boldsymbol{x}, s) = \mathbf{0}, \quad (\boldsymbol{x}, s) \in \mathscr{L} \times \mathbb{R}_{>0}, \tag{5.2}$$

$$\text{IC:} \quad \boldsymbol{u}_t(\boldsymbol{x}, 0) = \boldsymbol{u}_{t0}(\boldsymbol{x}), \quad \boldsymbol{x} \in \mathring{\mathscr{P}}_0. \tag{5.3}$$

The general solution to Eq. (5.1) has been derived in [2], and makes use of classic results from the well-established theory for linear PDEs [3, 4]. The main intuition behind the proposed solution is that Eq. (5.1) may be converted into a linear system of ODEs. Indeed, the change of coordinates $\boldsymbol{x} = \boldsymbol{x}(\boldsymbol{\rho}, \varsigma)$, $s = s(\boldsymbol{\rho}, \varsigma)$, $\boldsymbol{u}_t(\boldsymbol{x}, s) = \boldsymbol{u}_t(\boldsymbol{x}(\boldsymbol{\rho}, \varsigma), s(\boldsymbol{\rho}, \varsigma)) = \boldsymbol{\zeta}(\boldsymbol{\rho}, \varsigma)$ turns the PDEs (5.1) into

$$\frac{\mathrm{d}s(\boldsymbol{\rho}, \varsigma)}{\mathrm{d}\varsigma} = 1, \tag{5.4a}$$

$$\frac{\mathrm{d}\boldsymbol{x}(\boldsymbol{\rho}, \varsigma)}{\mathrm{d}\varsigma} = -\begin{bmatrix} 1 \\ 0 \end{bmatrix} + \mathbf{A}_{\varphi_\gamma}(s)\boldsymbol{x}(\boldsymbol{\rho}, \varsigma), \tag{5.4b}$$

$$\frac{\mathrm{d}\boldsymbol{\zeta}(\boldsymbol{\rho}, \varsigma)}{\mathrm{d}\varsigma} = \mathbf{A}_{\varphi_\psi}(s)\boldsymbol{\zeta}(\boldsymbol{\rho}, \varsigma) + \boldsymbol{\sigma}(s) + \mathbf{A}_{\varphi}(s)\boldsymbol{x}(\boldsymbol{\rho}, \varsigma), \tag{5.4c}$$

where $\mathbf{A}_{\varphi_\psi}(s) = \chi_\psi(s)\mathbf{A}_\varphi(s)$ is called *turning tensor* [2]. The system (5.4a), (5.4b), (5.4c) admits a solution in the form [2, 5]:

$$s(\boldsymbol{\rho}, \varsigma) = \varsigma + s_0(\boldsymbol{\rho}), \tag{5.5a}$$

$$\boldsymbol{x}(\boldsymbol{\rho}, \varsigma) = \mathbf{\Phi}_{\varphi_\gamma}(\varsigma, 0)\boldsymbol{x}_0(\boldsymbol{\rho}) - \int_0^{\varsigma} \mathbf{\Phi}_{\varphi_\gamma}(\varsigma, \varsigma') \begin{bmatrix} 1 \\ 0 \end{bmatrix} \mathrm{d}\varsigma', \tag{5.5b}$$

$$\boldsymbol{\zeta}(\boldsymbol{\rho}, \varsigma) = \mathbf{\Phi}_{\varphi_\psi}(\varsigma, 0)\boldsymbol{\zeta}_0(\boldsymbol{\rho}) + \int_0^{\varsigma} \mathbf{\Phi}_{\varphi_\psi}(\varsigma, \varsigma')\Big[\boldsymbol{\sigma}\big(\varsigma' + s_0(\boldsymbol{\rho})\big) + \mathbf{A}_{\varphi}\big(\varsigma' + s_0(\boldsymbol{\rho})\big)\boldsymbol{x}(\boldsymbol{\rho}, \varsigma')\Big] \mathrm{d}\varsigma', \tag{5.5c}$$

[1] As explained later, when φ_γ is constant, these trajectories become circles. In this case, it would be possible to solve the equations transforming into polar coordinates.

in which $\mathbf{\Phi}_{\varphi_\gamma}(\varsigma,\tilde{\varsigma})$ and $\mathbf{\Phi}_{\varphi_\psi}(\varsigma,\tilde{\varsigma})$ are called *camber* and *spin slip transition matrices*, respectively, and are given by [5]

$$\mathbf{\Phi}_{\varphi_\gamma}(\varsigma,\tilde{\varsigma}) = \mathrm{e}^{\int_{\tilde{\varsigma}}^{\varsigma}\mathbf{A}_{\varphi_\gamma}(\varsigma'+s_0(\boldsymbol{\rho}))\,\mathrm{d}\varsigma'} = \begin{bmatrix} \cos\Bigl(\int_{\tilde{\varsigma}}^{\varsigma}\varphi_\gamma(\varsigma'+s_0(\boldsymbol{\rho}))\,\mathrm{d}\varsigma'\Bigr) & \sin\Bigl(\int_{\tilde{\varsigma}}^{\varsigma}\varphi_\gamma(\varsigma'+s_0(\boldsymbol{\rho}))\,\mathrm{d}\varsigma'\Bigr) \\ -\sin\Bigl(\int_{\tilde{\varsigma}}^{\varsigma}\varphi_\gamma(\varsigma'+s_0(\boldsymbol{\rho}))\,\mathrm{d}\varsigma'\Bigr) & \cos\Bigl(\int_{\tilde{\varsigma}}^{\varsigma}\varphi_\gamma(\varsigma'+s_0(\boldsymbol{\rho}))\,\mathrm{d}\varsigma'\Bigr) \end{bmatrix}, \tag{5.6a}$$

$$\mathbf{\Phi}_{\varphi_\psi}(\varsigma,\tilde{\varsigma}) = \mathrm{e}^{\int_{\tilde{\varsigma}}^{\varsigma}\mathbf{A}_{\varphi_\psi}(\varsigma'+s_0(\boldsymbol{\rho}))\,\mathrm{d}\varsigma'} = \begin{bmatrix} \cos\Bigl(\int_{\tilde{\varsigma}}^{\varsigma}\varphi_\psi(\varsigma'+s_0(\boldsymbol{\rho}))\,\mathrm{d}\varsigma'\Bigr) & -\sin\Bigl(\int_{\tilde{\varsigma}}^{\varsigma}\varphi_\psi(\varsigma'+s_0(\boldsymbol{\rho}))\,\mathrm{d}\varsigma'\Bigr) \\ \sin\Bigl(\int_{\tilde{\varsigma}}^{\varsigma}\varphi_\psi(\varsigma'+s_0(\boldsymbol{\rho}))\,\mathrm{d}\varsigma'\Bigr) & \cos\Bigl(\int_{\tilde{\varsigma}}^{\varsigma}\varphi_\psi(\varsigma'+s_0(\boldsymbol{\rho}))\,\mathrm{d}\varsigma'\Bigr) \end{bmatrix}. \tag{5.6b}$$

The integral solutions displayed in Eqs. (5.5a), (5.5b), (5.5c) are always unique and well-posed, since the right-hand sides of Eqs. (5.4a), (5.4b), (5.4c) are globally Lipschitz. Accordingly, the solution to the original PDEs (5.1) may be found by inverting the mappings for $\boldsymbol{\rho}(\boldsymbol{x},s)$ and $\varsigma(\boldsymbol{x},s)$, yielding a relationship of the type $\boldsymbol{\zeta}(\boldsymbol{\rho}(\boldsymbol{x},s),\varsigma(\boldsymbol{x},s)) = \boldsymbol{u}_t(\boldsymbol{x},s)$. This operation is always possible if the BC or the IC are *noncharacteristic*, as better explained in Appendix 5.A. The noncharacteristic condition guarantees the existence of a C^2 function solving the PDEs (5.1) in the proximity of the boundary (or initial) curve [3, 4]. With respect to this specific problem, a proof of the noncharacteristic condition for the BC (5.2) and IC (5.3) was first given in [2] and is reported here in Appendix 5.A (Proposition 5.A.1).

In the classic transient brush theory presented in Chap. 4, however, it was already shown that often, even under vanishing sliding conditions, solutions only C^0 should be admitted owing to the nonanaliticity of the initial conditions (for example when $\boldsymbol{u}_{t0}(\boldsymbol{x})$ is only $C^0(\mathscr{P}_0)$). Furthermore, if the contact patch is not convex in the rolling direction $\bar{\boldsymbol{v}}_t(\boldsymbol{x})$, the global solution over $\mathscr{P}$ is discontinuous also in steady-state[2]. As clarified later in Subsect. 5.1.3, similar considerations are also valid for the exact theory discussed in this chapter. Furthermore, imposing the BC and IC provides, in turn, the stationary and transient solutions $\boldsymbol{u}_t^-(\boldsymbol{x},s)$ and $\boldsymbol{u}_t^+(\boldsymbol{x},s)$ for the deflection of the bristle. Accordingly, the contact patch itself may be separated into a stationary and transient region $\mathscr{P}^-$ and $\mathscr{P}^+$, respectively, with $\mathscr{P} = \mathscr{P}^- \cup \mathscr{P}^+$. In analogy to what done in Chap. 4, the global solution may be thus constructed as

$$\boldsymbol{u}_t(\boldsymbol{x},s) = \begin{cases} \boldsymbol{u}_t^-(\boldsymbol{x},s), & (\boldsymbol{x},s) \in \mathscr{P}^- \times \mathbb{R}_{\geq 0}, \\ \boldsymbol{u}_t^+(\boldsymbol{x},s), & (\boldsymbol{x},s) \in \mathscr{P}^+ \times \mathbb{R}_{\geq 0}. \end{cases} \tag{5.7}$$

The next Subsects. 5.1.1 and 5.1.2 are dedicated to the derivation of the analytical expressions for $\boldsymbol{u}_t^-(\boldsymbol{x},s)$ and $\boldsymbol{u}_t^+(\boldsymbol{x},s)$, respectively.

[2] The argument that the contact patch should be convex in the rolling direction is corroborated intuitively by Pacejka [6], and also justifies the adoption of the camber reduction factor ε_γ.

5.1.1 Steady-State Solution

Enforcing the BC (5.2) provides the stationary displacement of the bristle. In particular, a closed-form expression may be derived when considering a fixed contact geometry and constant slip inputs. In this condition, the stationary solution, which is also steady-state, is given by

$$s(\boldsymbol{\rho}, \varsigma) = \varsigma + s_0(\boldsymbol{\rho}), \tag{5.8a}$$

$$\boldsymbol{x}(\boldsymbol{\rho}, \varsigma) = \mathbf{R}_{\varphi_\gamma}(\varsigma)\big(\boldsymbol{x}_0(\boldsymbol{\rho}) - \boldsymbol{x}_{C_\gamma}\big) + \boldsymbol{x}_{C_\gamma}, \tag{5.8b}$$

$$\boldsymbol{\zeta}(\boldsymbol{\rho}, \varsigma) = \mathbf{R}_{\varphi_\psi}(\varsigma)\Big(\boldsymbol{\zeta}_0(\boldsymbol{\rho}) - \tilde{\boldsymbol{\zeta}}_0(\boldsymbol{\rho})\Big) + \tilde{\boldsymbol{\zeta}}(\boldsymbol{\rho}, \varsigma), \tag{5.8c}$$

since the transition matrices in Eqs. (5.6a), (5.6b) reduce to rotation matrices of the type [2]

$$\mathbf{R}_{\varphi_\gamma}(\varsigma) \triangleq \mathrm{e}^{\mathbf{A}_{\varphi_\gamma}\varsigma} = \begin{bmatrix} \cos(\varphi_\gamma\varsigma) & \sin(\varphi_\gamma\varsigma) \\ -\sin(\varphi_\gamma\varsigma) & \cos(\varphi_\gamma\varsigma) \end{bmatrix}, \tag{5.9a}$$

$$\mathbf{R}_{\varphi_\psi}(\varsigma) \triangleq \mathrm{e}^{\mathbf{A}_{\varphi_\psi}\varsigma} = \begin{bmatrix} \cos(\varphi_\psi\varsigma) & -\sin(\varphi_\psi\varsigma) \\ \sin(\varphi_\psi\varsigma) & \cos(\varphi_\psi\varsigma) \end{bmatrix}, \tag{5.9b}$$

and the vector $\boldsymbol{x}_{C_\gamma}$ in Eq. (5.8b), representing the coordinate of the *cambering centre* C_γ, reads specifically

$$\boldsymbol{x}_{C_\gamma} = \begin{bmatrix} 0 \ y_{C_\gamma} \end{bmatrix}^{\mathrm{T}} \triangleq \begin{bmatrix} 0 \ R_\gamma \end{bmatrix}^{\mathrm{T}} = \begin{bmatrix} 0 \ 1/\varphi_\gamma \end{bmatrix}^{\mathrm{T}}. \tag{5.10}$$

The function $\tilde{\boldsymbol{\zeta}}(\cdot, \cdot)$ appearing in Eq. (5.8c) is defined as [2]

$$\tilde{\boldsymbol{\zeta}}(\boldsymbol{\rho}, \varsigma) = -\mathbf{A}_{\varphi_\psi}^{-1}\boldsymbol{\sigma} - \boldsymbol{x}(\boldsymbol{\rho}, \varsigma) + \boldsymbol{x}_{C_\psi}, \tag{5.11}$$

with $\tilde{\boldsymbol{\zeta}}_0(\boldsymbol{\rho}) \triangleq \tilde{\boldsymbol{\zeta}}(\boldsymbol{\rho}, 0)$, and $\boldsymbol{x}_{C_\psi}$ denoting the position of the *turning centre* C_ψ:

$$\boldsymbol{x}_{C_\psi} = \begin{bmatrix} 0 \ y_{C_\psi} \end{bmatrix}^{\mathrm{T}} \triangleq \begin{bmatrix} 0 \ R_\psi \end{bmatrix}^{\mathrm{T}} = \begin{bmatrix} 0 \ -1/\varphi_\psi \end{bmatrix}^{\mathrm{T}}. \tag{5.12}$$

Inverting the mappings for $(\boldsymbol{\rho}, \varsigma)$ and $(\boldsymbol{x}, s)$, the steady-state deflection of the bristle may be found from Eqs. (5.8a), (5.8b), (5.8c) as $\boldsymbol{u}_t^-(\boldsymbol{x}, s) = \boldsymbol{\zeta}(\boldsymbol{\rho}(\boldsymbol{x}, s), \varsigma(\boldsymbol{x}, s))$. To this end, it is worth recalling that $\boldsymbol{\zeta}_0(\boldsymbol{\rho}) = \boldsymbol{0}$. Therefore, Eq. (5.8c) simplifies to

$$\boldsymbol{\zeta}(\boldsymbol{\rho}, \varsigma) = -\mathbf{R}_{\varphi_\psi}(\varsigma)\tilde{\boldsymbol{\zeta}}_0(\boldsymbol{\rho}) + \tilde{\boldsymbol{\zeta}}(\boldsymbol{\rho}, \varsigma). \tag{5.13}$$

Moreover, it may be noticed that $\tilde{\boldsymbol{\zeta}}(\boldsymbol{\rho}, \varsigma)$ is, in reality, already a function of the physical variables $\boldsymbol{x}$ through $\tilde{\boldsymbol{\zeta}}(\boldsymbol{\rho}(\boldsymbol{x}, s), \varsigma(\boldsymbol{x}, s)) = \tilde{\boldsymbol{u}}_t(\boldsymbol{x})$, with

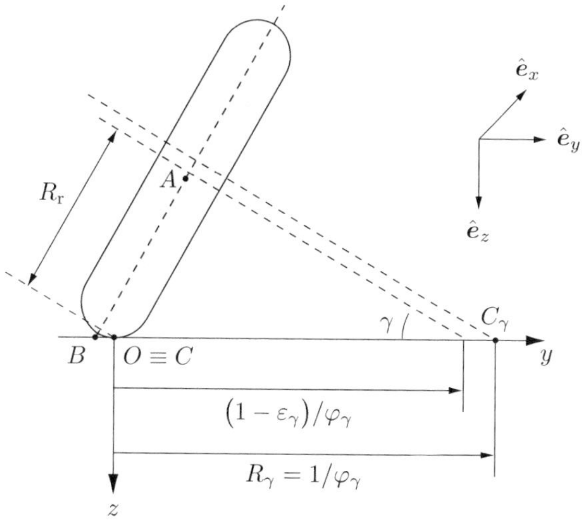

Fig. 5.1 For constant camber spin inputs, the trajectories of the bristles are circles centred in the cambering centre C_γ

$$\tilde{\boldsymbol{u}}_t(\boldsymbol{x}) \triangleq -\mathbf{A}_{\varphi_\psi}^{-1}\boldsymbol{\sigma} - \boldsymbol{x} + \boldsymbol{x}_{C_\psi}. \tag{5.14}$$

Additionally, manipulating Eqs. (5.8a) and (5.8b) leads to

$$x^2 + \left(y - 1/\varphi_\gamma\right)^2 = \left(x_0(\boldsymbol{\rho})\right)^2 + \left(y_0(\boldsymbol{\rho}) - 1/\varphi_\gamma\right)^2, \tag{5.15a}$$

$$\varsigma = \frac{1}{\varphi_\gamma}\left[\arctan\left(\frac{x}{y - 1/\varphi_\gamma}\right) - \arctan\left(\frac{x_0(\boldsymbol{\rho})}{y_0(\boldsymbol{\rho}) - 1/\varphi_\gamma}\right)\right], \tag{5.15b}$$

$$s_0(\boldsymbol{\rho}) = s - \varsigma, \tag{5.15c}$$

which allows expressing the space coordinates $\boldsymbol{x}$ as a sole function of $\boldsymbol{\rho}$. This result should be expected, and has the physical meaning that the characteristic projections (basically, the trajectories of the bristles travelling inside the contact patch) are circles of arbitrary radius centred in the cambering centre[3] C_γ (Fig. 5.1). According to Eq. (5.15a), the radius of these circles is also constant over time or, equivalently, the travelled distance s [2, 5, 7]. In light of the previous considerations, it should be clear that introducing the camber reduction factor ε_γ has the effect of increasing the radius of curvature $|R_\gamma|$ of the path of the bristles.

The solution $\boldsymbol{u}_t^-(\boldsymbol{x}, s)$ applies to the steady-state region of the contact patch $\mathscr{P}^-$, and may be found combining Eq. (5.8b) with (5.15a), (5.15b), (5.15c). For a fixed geometry, Eqs. (5.15a), (5.15b), (5.15c) may be inverted by setting $s_0(\boldsymbol{\rho}) = \rho_1$ and $\boldsymbol{x}_0(\boldsymbol{\rho}) = \boldsymbol{x}_0(\rho_2)$, which yields

$$\boldsymbol{u}_t^-(\boldsymbol{x}) = \mathbf{R}_{\varphi_\psi}\left(\Sigma(\boldsymbol{x})\right)\boldsymbol{\Psi}(\boldsymbol{x}) + \tilde{\boldsymbol{u}}_t(\boldsymbol{x}), \qquad (\boldsymbol{x}, s) \in \mathscr{P}^- \times \mathbb{R}_{\geq 0}, \tag{5.16}$$

[3] This happens because the velocity field $\bar{\boldsymbol{v}}_t(\boldsymbol{x})$ is solenoidal [2], i.e. $\nabla_t \cdot \bar{\boldsymbol{v}}_t(\boldsymbol{x}) = 0$.

in which $\Sigma(\cdot)$ and $\boldsymbol{\Psi}(\cdot)$ read [2]

$$\Sigma(\boldsymbol{x}) \triangleq \frac{1}{\varphi_\gamma}\left[\arctan\left(\frac{x}{y-1/\varphi_\gamma}\right) - \arctan\left(\frac{x_0\big(\rho_2(\boldsymbol{x})\big)}{y_0\big(\rho_2(\boldsymbol{x})\big)-1/\varphi_\gamma}\right)\right], \tag{5.17a}$$

$$\boldsymbol{\Psi}(\boldsymbol{x}) = \Big[\Psi_x(\boldsymbol{x})\ \Psi_y(\boldsymbol{x})\Big]^{\mathrm{T}} \triangleq -\tilde{\boldsymbol{u}}_t\Big(\boldsymbol{x}_0\big(\rho_2(\boldsymbol{x})\big)\Big). \tag{5.17b}$$

The expression for the steady-state region of the contact patch $\mathscr{P}^-$ in Eq. (5.16) may be found by imposing $s_0(\rho_1(\boldsymbol{x},s)) = \rho_1(\boldsymbol{x},s) > 0$ in Eq. (5.15c). Recalling (5.15b), this is equivalent to having $s > \Sigma(\boldsymbol{x})$, yielding

$$\mathscr{P}^- \triangleq \Big\{\boldsymbol{x} \in \mathscr{P} \ \Big|\ \gamma_\Sigma(\boldsymbol{x},s) < 0\Big\}, \tag{5.18}$$

where $\gamma_\Sigma(\boldsymbol{x},s) = 0$ describes again the travelling edge, with

$$\gamma_\Sigma(\boldsymbol{x},s) \triangleq \Sigma(\boldsymbol{x}) - s. \tag{5.19}$$

From Eq. (5.16), it may be inferred that the analytical expression for the steady-state deflection $\boldsymbol{u}_t^-(\boldsymbol{x})$ depends explicitly upon the shape of the leading edge, but not on the travelled distance s. This is a direct consequence of the assumption of constant contact geometry and slip inputs. On the other hand, when these are also allowed to vary over time, the solution may still be interpreted as a stationary one, since it persists after the initial transient phase, but is also a function of s. This is in accordance with what has been found already in Chap. 4.

Moreover, similar to what has been done in Chap. 4, the dynamics of the travelling edge may be studied starting from the analytical expression of the unit normal

$$\hat{\boldsymbol{\nu}}_\Sigma(\boldsymbol{x}) = \pm\frac{\nabla_t\Sigma(\boldsymbol{x})}{\big\|\nabla_t\Sigma(\boldsymbol{x})\big\|}. \tag{5.20}$$

Accordingly, a particular representation of the velocity of the travelling edge that is oriented as the unit normal reads

$$\bar{\boldsymbol{v}}_\Sigma^{(\hat{\nu})}(\boldsymbol{x}) = \frac{\nabla_t\Sigma(\boldsymbol{x})}{\big\|\nabla_t\Sigma(\boldsymbol{x})\big\|^2}, \tag{5.21}$$

being clearly $\partial\gamma_\Sigma(\boldsymbol{x},s)/\partial s = -1$.

5.1.2 *Transient Solution*

The IC (5.3) enforces $s_0(\boldsymbol{\rho}) = 0$. Therefore, the transient solution $\boldsymbol{u}_t^+(\boldsymbol{x},s)$ may be easily derived by setting $\boldsymbol{x}_0(\boldsymbol{\rho}) = \boldsymbol{\rho}$ and inverting for $\boldsymbol{\rho}$. In particular, the compatibility condition $\boldsymbol{\zeta}_0(\boldsymbol{\rho}(\boldsymbol{x},s)) = \boldsymbol{u}_{t0}(\boldsymbol{x}_0(\boldsymbol{x},s))$ yields [2]

$$u_t^+(x,s) = \Phi_{\varphi_\psi}(s,0)u_{t0}\big(x_0(x,s)\big) + \int_0^s \Phi_{\varphi_\psi}(s,s')\Big[\sigma(s') + \mathbf{A}_\varphi(s')x(\rho,s')\Big]\,\mathrm{d}s', \tag{5.22}$$

where $x_0(x,s)$ is given by

$$x_0(x,s) = \Phi_{\varphi_\gamma}(0,s)\left[x + \int_0^s \Phi_{\varphi_\gamma}(s,s')\begin{bmatrix}1\\0\end{bmatrix}\mathrm{d}s'\right]. \tag{5.23}$$

Additionally, assuming constant slip inputs, Eqs. (5.22) and (5.23) reduce to [2]

$$u_t^+(x,s) = \mathbf{R}_{\varphi_\psi}(s)\Big[u_{t0}\big(x_0(x,s)\big) - \tilde{u}_{t0}(x,s)\Big] + \tilde{u}_t(x), \quad (x,s) \in \mathscr{P}^+ \times \mathbb{R}_{\geq 0}, \tag{5.24}$$

with $\tilde{u}_{t0}(x,s) \triangleq \tilde{u}_t(x_0(x,s))$, and

$$x_0(x,s) = \rho(x,s) = \mathbf{R}_{\varphi_\gamma}^{-1}(s)\big(x - x_{C_\gamma}\big) + x_{C_\gamma}. \tag{5.25}$$

In the latter case, the transient region of the contact patch $\mathscr{P}^+$ is given by

$$\mathscr{P}^+ \triangleq \big\{x \in \mathscr{P} \mid \gamma_\Sigma(x,s) \geq 0\big\}, \tag{5.26}$$

and the solution appears to be continuous on the travelling edge, since it holds identically that $u_t^+(x,\Sigma(x)) = u_t^-(x)$.

From both Eqs. (5.22) and (5.24), it may be deduced that, as opposed to the steady-state deflection of the bristle, the transient solution does not depend directly upon the contact shape. Indeed, the inversion formula given by Eq. (5.24) is valid for any geometry.

5.1.3 Explicit Solutions for Some Contact Shapes

Analytical solutions for the steady-state deflection of the bristle have been derived in [2] for rectangular and elliptical contact shapes, owing to the conditions

$$y < R_\gamma = \frac{1}{|\varphi_\gamma|}, \qquad \varphi_\gamma > 0, \qquad x \in \mathscr{P}, \tag{5.27a}$$

$$y > R_\gamma = -\frac{1}{|\varphi_\gamma|}, \qquad \varphi_\gamma < 0, \qquad x \in \mathscr{P}. \tag{5.27b}$$

The inequalities (5.27a), (5.27b) are always satisfied in reality, since it follows from geometrical considerations that

$$y < \frac{R_{\mathrm{r}}}{|\sin\gamma|} \leq \frac{R_{\mathrm{r}}}{(1-\varepsilon_\gamma)|\sin\gamma|} = \frac{1}{|\varphi_\gamma|} = R_\gamma, \qquad \varphi_\gamma > 0, \qquad \boldsymbol{x} \in \mathscr{P}, \tag{5.28a}$$

$$y > -\frac{R_{\mathrm{r}}}{|\sin\gamma|} \geq -\frac{R_{\mathrm{r}}}{(1-\varepsilon_\gamma)|\sin\gamma|} = -\frac{1}{|\varphi_\gamma|} = R_\gamma, \quad \varphi_\gamma < 0, \quad \boldsymbol{x} \in \mathscr{P}. \tag{5.28b}$$

5.1.3.1 Rectangular Contact Patch

When the velocity field is as in Eq. (2.12), a rectangular patch is not D-convex in the rolling direction $\bar{\boldsymbol{v}}_t(\boldsymbol{x})$, and therefore the leading edge cannot be parametrised by a continuous smooth function. As a result, even in steady-state conditions, no solution that is globally $C^0(\mathscr{P})$ can be found. Indeed, three different BCs, which depend upon the sign of the camber angle γ and thus upon φ_γ, needs to be prescribed. The corresponding expressions for the deflection of the bristles are denoted in the following by $\boldsymbol{u}_{1t}^-(\boldsymbol{x})$, $\boldsymbol{u}_{2t}^-(\boldsymbol{x})$ and $\boldsymbol{u}_{3t}^-(\boldsymbol{x})$. In particular, the contact patch is defined mathematically as

$$\mathscr{P} \triangleq \big\{\boldsymbol{x} \in \Pi \mid -a \leq x \leq a,\ -b \leq y \leq b\big\}, \tag{5.29}$$

with the three leading edges parametrised by

$$x = x_{\mathscr{L}_1}(y) = a, \qquad y \in (-b, b), \tag{5.30a}$$
$$y = y_{\mathscr{L}_2}(x) = b \operatorname{sgn} \varphi_\gamma, \qquad x \in (0, a), \tag{5.30b}$$
$$y = y_{\mathscr{L}_3}(x) = -b \operatorname{sgn} \varphi_\gamma, \qquad x \in (-a, 0). \tag{5.30c}$$

The contact patch may be thus partitioned as

$$\mathscr{P}_1 \triangleq \big\{\mathscr{P} \setminus (\mathscr{P}_2 \cup \mathscr{P}_3)\big\}, \tag{5.31a}$$
$$\mathscr{P}_2 \triangleq \Big\{\boldsymbol{x} \in \mathscr{P} \;\Big|\; R_1^2 < \Gamma(\boldsymbol{x}) < R_2^2\Big\}, \tag{5.31b}$$
$$\mathscr{P}_3 \triangleq \Big\{\boldsymbol{x} \in \mathscr{P} \;\Big|\; R_3^2 < \Gamma(\boldsymbol{x}) < R_4^2,\ x < 0\Big\}, \tag{5.31c}$$

with $\Gamma(\boldsymbol{x})$ reading

$$\Gamma(\boldsymbol{x}) \triangleq x^2 + \big(y - 1/\varphi_\gamma\big)^2, \tag{5.32}$$

and the radii R_0, R_1, R_2, R_3 and R_4 defined as

$$R_0 \triangleq a, \qquad R_1 \triangleq b \operatorname{sgn} \varphi_\gamma - 1/\varphi_\gamma, \qquad R_2 \triangleq \sqrt{R_1^2 + R_0^2},$$
$$R_3 \triangleq b \operatorname{sgn} \varphi_\gamma + 1/\varphi_\gamma, \qquad R_4 \triangleq \sqrt{R_3^2 + R_0^2}. \tag{5.33}$$

The above radii correspond to the following circles:

$$\mathcal{C}_0(\boldsymbol{x}) \triangleq \left\{\boldsymbol{x} \in \Pi \;\middle|\; \Gamma(\boldsymbol{x}) = R_0^2\right\}, \tag{5.34a}$$

$$\mathcal{C}_1(\boldsymbol{x}) \triangleq \left\{\boldsymbol{x} \in \Pi \;\middle|\; \Gamma(\boldsymbol{x}) = R_1^2\right\}, \tag{5.34b}$$

$$\mathcal{C}_2(\boldsymbol{x}) \triangleq \left\{\boldsymbol{x} \in \Pi \;\middle|\; \Gamma(\boldsymbol{x}) = R_2^2\right\}, \tag{5.34c}$$

$$\mathcal{C}_3(\boldsymbol{x}) \triangleq \left\{\boldsymbol{x} \in \Pi \;\middle|\; \Gamma(\boldsymbol{x}) = R_3^2\right\}, \tag{5.34d}$$

$$\mathcal{C}_4(\boldsymbol{x}) \triangleq \left\{\boldsymbol{x} \in \Pi \;\middle|\; \Gamma(\boldsymbol{x}) = R_4^2\right\}. \tag{5.34e}$$

Each of the regions in Eq. (5.31a), (5.31b), (5.31c) may be further divided into a steady-state subdomain $\mathscr{P}_i^-$, and a transient one $\mathscr{P}_i^+$:

$$\mathscr{P}_1^- \triangleq \left\{\boldsymbol{x} \in \mathscr{P}_1 \mid \gamma_{\Sigma_1}(\boldsymbol{x}, s) < 0\right\}, \quad \mathscr{P}_1^+ \triangleq \left\{\boldsymbol{x} \in \mathscr{P}_1 \mid \gamma_{\Sigma_1}(\boldsymbol{x}, s) \geq 0\right\}, \tag{5.35a}$$

$$\mathscr{P}_2^- \triangleq \left\{\boldsymbol{x} \in \mathscr{P}_2 \mid \gamma_{\Sigma_2}(\boldsymbol{x}, s) < 0\right\}, \quad \mathscr{P}_2^+ \triangleq \left\{\boldsymbol{x} \in \mathscr{P}_2 \mid \gamma_{\Sigma_2}(\boldsymbol{x}, s) \geq 0\right\}, \tag{5.35b}$$

$$\mathscr{P}_3^- \triangleq \left\{\boldsymbol{x} \in \mathscr{P}_3 \mid \gamma_{\Sigma_3}(\boldsymbol{x}, s) < 0\right\}, \quad \mathscr{P}_3^+ \triangleq \left\{\boldsymbol{x} \in \mathscr{P}_3 \mid \gamma_{\Sigma_3}(\boldsymbol{x}, s) \geq 0\right\}. \tag{5.35c}$$

The corresponding steady-state solutions read

$$\boldsymbol{u}_{1t}^-(\boldsymbol{x}) = \mathbf{R}_{\varphi_\psi}\big(\Sigma_1(\boldsymbol{x})\big)\boldsymbol{\Psi}_1(\boldsymbol{x}) + \tilde{\boldsymbol{u}}_t(\boldsymbol{x}), \qquad (\boldsymbol{x}, s) \in \mathscr{P}_1^- \times \mathbb{R}_{\geq 0}, \tag{5.36a}$$

$$\boldsymbol{u}_{2t}^-(\boldsymbol{x}) = \mathbf{R}_{\varphi_\psi}\big(\Sigma_2(\boldsymbol{x})\big)\boldsymbol{\Psi}_2(\boldsymbol{x}) + \tilde{\boldsymbol{u}}_t(\boldsymbol{x}), \qquad (\boldsymbol{x}, s) \in \mathscr{P}_2^- \times \mathbb{R}_{\geq 0}, \tag{5.36b}$$

$$\boldsymbol{u}_{3t}^-(\boldsymbol{x}) = \mathbf{R}_{\varphi_\psi}\big(\Sigma_3(\boldsymbol{x})\big)\boldsymbol{\Psi}_3(\boldsymbol{x}) + \tilde{\boldsymbol{u}}_t(\boldsymbol{x}), \qquad (\boldsymbol{x}, s) \in \mathscr{P}_3^- \times \mathbb{R}_{\geq 0}, \tag{5.36c}$$

where the functions $\Sigma_1(\cdot)$, $\Sigma_2(\cdot)$ and $\Sigma_3(\cdot)$ are given by

$$\Sigma_1(\boldsymbol{x}) \triangleq \frac{1}{\varphi_\gamma}\left[\arctan\left(\frac{x}{y - 1/\varphi_\gamma}\right) + \arctan\left(\frac{R_0}{\sqrt{\Gamma(\boldsymbol{x}) - R_0^2}}\right)\operatorname{sgn}\varphi_\gamma\right], \tag{5.37a}$$

$$\Sigma_2(\boldsymbol{x}) \triangleq \frac{1}{\varphi_\gamma}\left[\arctan\left(\frac{x}{y - 1/\varphi_\gamma}\right) - \arctan\left(\frac{\sqrt{\Gamma(\boldsymbol{x}) - R_1^2}}{R_1}\right)\right], \tag{5.37b}$$

$$\Sigma_3(\boldsymbol{x}) \triangleq \frac{1}{\varphi_\gamma}\left[\arctan\left(\frac{x}{y - 1/\varphi_\gamma}\right) - \arctan\left(\frac{\sqrt{\Gamma(\boldsymbol{x}) - R_3^2}}{R_3}\right)\right], \tag{5.37c}$$

and $\boldsymbol{\Psi}_1(\cdot)$, $\boldsymbol{\Psi}_2(\cdot)$, $\boldsymbol{\Psi}_3(\cdot)$ write in components

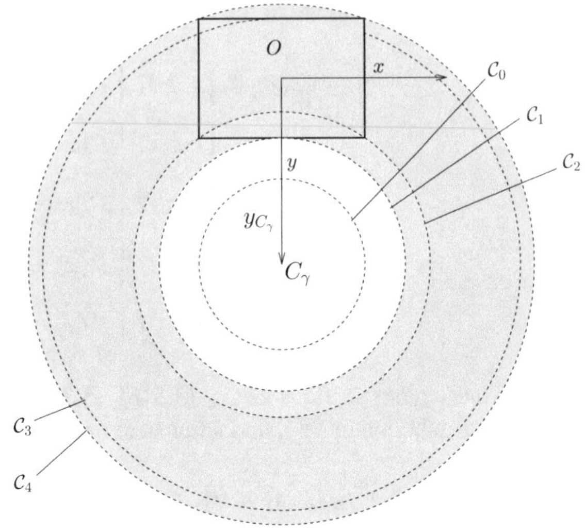

Fig. 5.2 Domains of the steady-state solutions for a rectangular patch. The green, red and blue areas inside the rectangle coincide with the domains $\mathscr{P}_1$, $\mathscr{P}_2$ and $\mathscr{P}_3$, where the corresponding solutions take place. The displacement of the bristle is discontinuous on the circle $\mathcal{C}_3(\boldsymbol{x})$

$$\Psi_{1x}(\boldsymbol{x}) \triangleq \frac{\sigma_y}{\varphi_\psi} + a, \tag{5.38a}$$

$$\Psi_{2x}(\boldsymbol{x}) \triangleq \frac{\sigma_y}{\varphi_\psi} + \sqrt{\Gamma(\boldsymbol{x}) - R_1^2}, \tag{5.38b}$$

$$\Psi_{3x}(\boldsymbol{x}) \triangleq \frac{\sigma_y}{\varphi_\psi} - \sqrt{\Gamma(\boldsymbol{x}) - R_3^2}, \tag{5.38c}$$

$$\Psi_{1y}(\boldsymbol{x}) \triangleq \frac{1}{\varphi_\psi}(1-\sigma_x) - \sqrt{\Gamma(\boldsymbol{x}) - R_0^2}\,\mathrm{sgn}\,\varphi_\gamma + \frac{1}{\varphi_\gamma}, \tag{5.38d}$$

$$\Psi_{2y}(\boldsymbol{x}) \triangleq \frac{1}{\varphi_\psi}(1-\sigma_x) + b\,\mathrm{sgn}\,\varphi_\gamma, \tag{5.38e}$$

$$\Psi_{3y}(\boldsymbol{x}) \triangleq \frac{1}{\varphi_\psi}(1-\sigma_x) - b\,\mathrm{sgn}\,\varphi_\gamma. \tag{5.38f}$$

The global solution constructed over $\mathscr{P}$ is neither $C^1(\mathring{\mathscr{P}} \times \mathbb{R}_{>0}; \mathbb{R}^2)$ nor $C^0(\mathscr{P} \times \mathbb{R}_{\geq 0}; \mathbb{R}^2)$, since the deflection of the bristle is discontinuous on $\mathcal{C}_3(\boldsymbol{x})$. On the other hand, the continuity is preserved in the transition between $\mathscr{P}_1$ and $\mathscr{P}_2$. This result is a direct consequence of the fact that the deflection is continuous on the curve $x = x_{\mathscr{L}_1}(y) = a$, as also illustrated in Fig. 5.2.

5.1.3.2 Elliptical Contact Patch

For an elliptical contact patch, the number of possible solutions depends on the magnitude of camber radius $|R_\gamma| = 1/|\varphi_\gamma|$: if this is too small compared to the ratio a/b between the semilength and width of the contact patch, multiple solutions arise that may be discontinuous in space, as it happens for the rectangular shape. If the inequality $a^2 \leq b(b + 1/|\varphi_\gamma|)$ is verified, a unique solution $\boldsymbol{u}_t^-(\boldsymbol{x}) \in C^1(\mathring{\mathscr{P}}^- \times \mathbb{R}_{>0}; \mathbb{R}^2)$ may always be found. Indeed, a contact patch for which the above condition is fulfilled is D-convex in the rolling direction $\bar{\boldsymbol{v}}_t(\boldsymbol{x})$, and may be described mathematically as

$$\mathscr{P} \triangleq \left\{ \boldsymbol{x} \in \Pi \;\middle|\; \frac{x^2}{a^2} + \frac{y^2}{b^2} \leq 1 \right\}. \tag{5.39}$$

In this case, the leading and trailing edges may be again parametrised as

$$x = x_{\mathscr{L}}(y) = a\sqrt{1 - \frac{y^2}{b^2}}, \qquad y \in (-b, b), \tag{5.40a}$$

$$x = x_{\mathscr{T}}(y) = -a\sqrt{1 - \frac{y^2}{b^2}}, \qquad y \in (-b, b). \tag{5.40b}$$

The solution may be derived by defining the radii

$$R_1 \triangleq b \operatorname{sgn} \varphi_\gamma - 1/\varphi_\gamma, \qquad R_2 \triangleq b \operatorname{sgn} \varphi_\gamma + 1/\varphi_\gamma, \tag{5.41}$$

and the corresponding circles

$$\mathcal{C}_1(\boldsymbol{x}) \triangleq \left\{ \boldsymbol{x} \in \Pi \;\middle|\; \Gamma(\boldsymbol{x}) = R_1^2 \right\}, \tag{5.42a}$$

$$\mathcal{C}_2(\boldsymbol{x}) \triangleq \left\{ \boldsymbol{x} \in \Pi \;\middle|\; \Gamma(\boldsymbol{x}) = R_2^2 \right\}, \tag{5.42b}$$

which are tangent to $\mathscr{P}$ at the neutral points $\boldsymbol{x}_{\mathscr{N}_1} = (0, b \operatorname{sgn} \varphi_\gamma)$ and $\boldsymbol{x}_{\mathscr{N}_2} = (0, -b \operatorname{sgn} \varphi_\gamma)$, respectively. In Eqs. (5.42a), (5.42b), the function $\Gamma(\cdot)$ is the same as that in Eq. (5.32).

Once again, the global solution over $\mathscr{P}$ may be constructed with the aid of Eqs. (5.15a), (5.15b), (5.15c) and Eq. (5.13). Specifically, for the steady-state deflection in $\mathscr{P}^-$, the following expression may be derived for $y_0(\rho_2(\boldsymbol{x}))$:

$$y_0\big(\rho_2(\boldsymbol{x})\big) = \frac{\dfrac{1}{\varphi_\gamma} - \sqrt{\dfrac{1}{\varphi_\gamma^2} - \left(1 - \dfrac{a^2}{b^2}\right)\left(\dfrac{1}{\varphi_\gamma^2} + a^2 - \Gamma(\boldsymbol{x})\right)} \operatorname{sgn} \varphi_\gamma}{\left(1 - \dfrac{a^2}{b^2}\right)}, \tag{5.43}$$

and $x_0(\rho_2(\boldsymbol{x})) = x_{\mathscr{L}} \circ y_0(\rho_2(\boldsymbol{x}))$ with $x_{\mathscr{L}}(\cdot)$ as in Eq. (5.40a).

5.2 Limited Friction

The case of limited friction is analysed only qualitatively for the steady-state case. The adhesion solutions may be obtained analytically as explained in the preceding sections, whilst the displacement of the bristle and the shear stress in the sliding zone is given by Eq. (2.2) and need to be calculated numerically, for example assuming a parabolic pressure distribution. Other simplifications may be also needed. As a first approximation, the partial derivatives of the bristle displacement may be for instance disregarded in the computation of the micro-siding velocity [6].

Figure 5.3 compares the friction circle and the trend of self-aligning moment versus the longitudinal slip for the exact and classic brush theory for a rectangular contact patch with $a = 0.075$ and $b = 0.05$ m. Both plots refer to a subcritical spin value of $\varphi = 3.33 < \varphi^{\text{cr}} = 4.00\ \text{m}^{-1}$, in combination with a camber ratio of $\chi_\gamma = 0.9$. The vertical force is set to $F_z = 3000$ N. As usual, a tyre with isotropic tread is assumed, with bristle stiffness $k = k_{xx} = k_{yy} = 2.67 \cdot 10^7$ and $k_{xy} = k_{yx} = 0\ \text{N}\,\text{m}^{-3}$. In Fig. 5.3, the solid lines refer to the forces and moment calculated according to the classic theory, whilst the dashed lines to those predicted by the exact one. A light discrepancy between the two formulations may be highlighted especially at lower values of the translational slips, where the camber spin appears to exert a greater influence upon the generation of tyre characteristics.

A similar comparison is also shown for an elliptical contact patch in Fig. 5.4 with the same longitudinal and lateral dimensions. If all the other parameters are fixed, the corresponding bristle stiffness reads $k = 4.52 \cdot 10^7\ \text{N}\,\text{m}^{-3}$. In this case, a better agreement is found between the classic and exact theories. This should be ascribed to the specific shape of the contact patch.

5.A Statement and Proof of Proposition 5.A.1

Proposition 5.A.1 *The BC* (5.2) *and IC* (5.3) *are noncharacteristic for the PDEs* (5.1).

Proof. Consider the Jacobian matrix

$$\mathbf{J}(\boldsymbol{\rho}, \varsigma) = \begin{bmatrix} \dfrac{\partial x(\boldsymbol{\rho}, \varsigma)}{\partial \rho_1} & \dfrac{\partial x(\boldsymbol{\rho}, \varsigma)}{\partial \rho_2} & \dfrac{\partial x(\boldsymbol{\rho}, \varsigma)}{\partial \varsigma} \\ \dfrac{\partial y(\boldsymbol{\rho}, \varsigma)}{\partial \rho_1} & \dfrac{\partial y(\boldsymbol{\rho}, \varsigma)}{\partial \rho_2} & \dfrac{\partial y(\boldsymbol{\rho}, \varsigma)}{\partial \varsigma} \\ \dfrac{\partial s(\boldsymbol{\rho}, \varsigma)}{\partial \rho_1} & \dfrac{\partial s(\boldsymbol{\rho}, \varsigma)}{\partial \rho_2} & \dfrac{\partial s(\boldsymbol{\rho}, \varsigma)}{\partial \varsigma} \end{bmatrix}. \tag{5.44}$$

Then, the BC and IC are said to be noncharacteristic if $\det \mathbf{J}(\boldsymbol{\rho}, 0) \neq 0, \infty$. Thus, it should be proved that the determinant of the Jacobian matrix in Eq. (5.44) evaluated at $\varsigma = 0$ never vanishes nor diverges when imposing the BC (5.2) and IC (5.3).

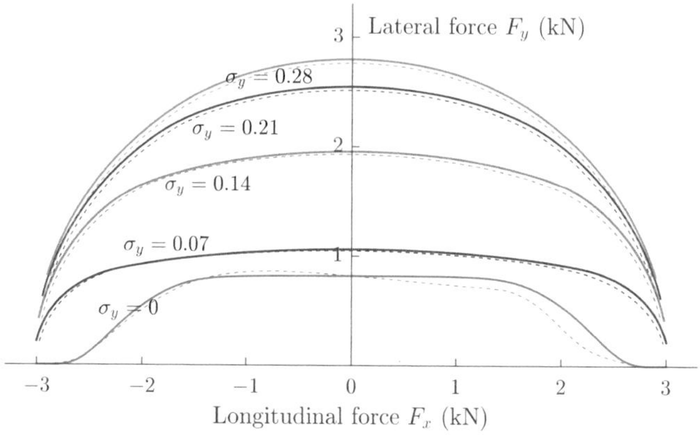

(a) Lateral force F_y versus longitudinal tyre force F_x.

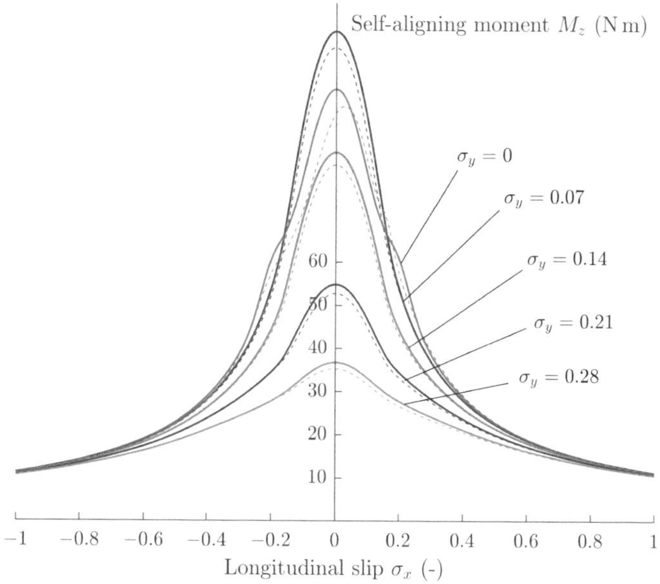

(b) Self-aligning moment M_z versus longitudinal slip σ_x.

Fig. 5.3 Steady-state characteristics predicted by the classic (solid lines) and exact (dashed lines) theories for different values of the lateral slip σ_y for a rectangular contact patch. Tyre parameters: $\varphi = 3.33\ \text{m}^{-1}$, $\chi_\gamma = 0.9$, $F_z = 3000$ N, $a = 0.075$ m, $b = 0.05$ m, $\mu = 1$

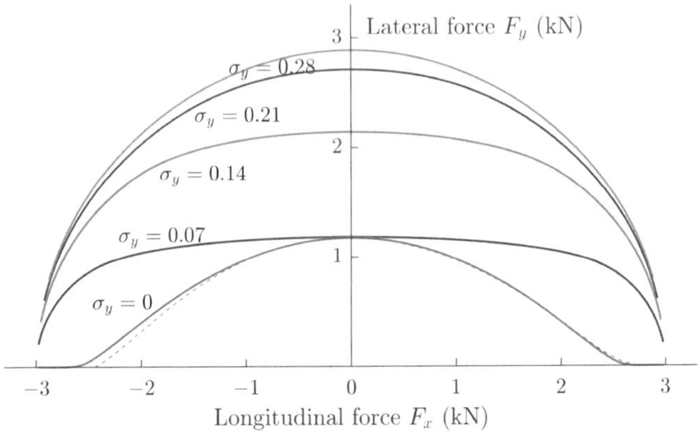

(a) Lateral force F_y versus longitudinal tyre force F_x.

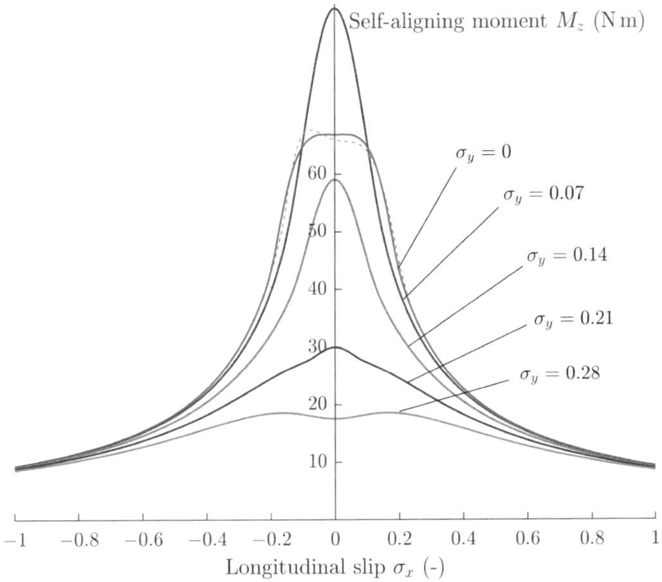

(b) Self-aligning moment M_z versus longitudinal slip σ_x.

Fig. 5.4 Steady-state characteristics predicted by the classic (solid lines) and exact (dashed lines) theories for different values of the lateral slip σ_y for an elliptical contact patch. Tyre parameters: $\varphi = 3.33$ m^{-1}, $\chi_\gamma = 0.9$, $F_z = 3000$ N, $a = 0.075$ m, $b = 0.05$ m, $\mu = 1$

The problem is first analysed when imposing the BC. In this case, on the boundary curve $\varsigma = 0$, the initial coordinates are given by $\boldsymbol{x}_0(\boldsymbol{\rho}) = \boldsymbol{x}_{\mathscr{L}}(\boldsymbol{\rho})$, where $\boldsymbol{x}_{\mathscr{L}}(\boldsymbol{\rho})$ is a local parametrisation of the leading edge. On the other hand, the initial conditions for the travelled distance may be chosen, without loss of generality, as $s_0(\boldsymbol{\rho}) = \rho_1$. Therefore, the Jacobian matrix in (5.44) becomes

$$\mathbf{J}_0\big(\boldsymbol{x}_0(\boldsymbol{\rho}), s_0(\rho_1)\big) = \begin{bmatrix} \dfrac{\partial x_0(\boldsymbol{\rho})}{\partial \rho_1} & \dfrac{\partial x_0(\boldsymbol{\rho})}{\partial \rho_2} & \bar{v}_x\big(y_0(\boldsymbol{\rho}), \rho_1\big) \\ \dfrac{\partial y_0(\boldsymbol{\rho})}{\partial \rho_1} & \dfrac{\partial y_0(\boldsymbol{\rho})}{\partial \rho_2} & \bar{v}_y\big(x_0(\boldsymbol{\rho}), \rho_1\big) \\ 1 & 0 & 1 \end{bmatrix}, \tag{5.45}$$

and its determinant may be written succinctly as

$$\det \mathbf{J}_0\big(\boldsymbol{x}_0(\boldsymbol{\rho}), s_0(\rho_1)\big) = \Big[\bar{\boldsymbol{v}}_t\big(\boldsymbol{x}_0(\boldsymbol{\rho}), \rho_1\big) - \bar{\boldsymbol{v}}_{\mathscr{L}}\big(\boldsymbol{x}_0(\boldsymbol{\rho})\big)\Big] \cdot \boldsymbol{\nu}_{\mathscr{L}}\big(\boldsymbol{x}_0(\boldsymbol{\rho})\big), \tag{5.46}$$

being

$$\boldsymbol{\nu}_{\mathscr{L}}\big(\boldsymbol{x}_0(\boldsymbol{\rho})\big) \triangleq \begin{bmatrix} -\dfrac{\partial y_0(\boldsymbol{\rho})}{\partial \rho_2} & \dfrac{\partial x_0(\boldsymbol{\rho})}{\partial \rho_2} \end{bmatrix}^{\mathrm{T}}, \tag{5.47a}$$

$$\bar{\boldsymbol{v}}_{\mathscr{L}}\big(\boldsymbol{x}_0(\boldsymbol{\rho})\big) \triangleq \begin{bmatrix} \dfrac{\partial x_0(\boldsymbol{\rho})}{\partial \rho_1} & \dfrac{\partial y_0(\boldsymbol{\rho})}{\partial \rho_1} \end{bmatrix}^{\mathrm{T}}. \tag{5.47b}$$

The above result may be better interpreted by recalling that, at $\varsigma = 0$, $s \equiv s_0(\boldsymbol{\rho}) = \rho_1$, and therefore the vector $\bar{\boldsymbol{v}}_{\mathscr{L}}(\boldsymbol{x}_0(\boldsymbol{\rho}))$ actually represents the nondimensional velocity of leading edge. Equations (5.46) and (5.47a), (5.47b) basically state that the bristles entering the contact patch cannot have the same velocity of the leading edge. The unit normal $\hat{\boldsymbol{\nu}}_{\mathscr{L}}(\boldsymbol{x}_0(\boldsymbol{\rho}))$ at $\boldsymbol{x}_0(\boldsymbol{\rho})$, if existing, is given by

$$\hat{\boldsymbol{\nu}}_{\mathscr{L}}\big(\boldsymbol{x}_0(\boldsymbol{\rho})\big) = \pm \frac{\boldsymbol{\nu}_{\mathscr{L}}\big(\boldsymbol{x}_0(\boldsymbol{\rho})\big)}{\Big\|\boldsymbol{\nu}_{\mathscr{L}}\big(\boldsymbol{x}_0(\boldsymbol{\rho})\big)\Big\|} = \pm \frac{\boldsymbol{\nu}_{\mathscr{L}}\big(\boldsymbol{x}_0(\boldsymbol{\rho})\big)}{\sqrt{\left(\dfrac{\partial x_0(\boldsymbol{\rho})}{\partial \rho_2}\right)^2 + \left(\dfrac{\partial y_0(\boldsymbol{\rho})}{\partial \rho_2}\right)^2}}. \tag{5.48}$$

Since $\bar{\boldsymbol{v}}_t(\boldsymbol{x}, s)$ is Lipschitz and thus bounded in $\mathscr{P}$, combining Eqs. (5.46) and (5.48) yields

$$\det \mathbf{J}_0\big(\boldsymbol{x}_0(\boldsymbol{\rho}), s_0(\rho_1)\big) = \pm\Big[\bar{\boldsymbol{v}}_t\big(\boldsymbol{x}_0(\boldsymbol{\rho}), \rho_1\big) - \bar{\boldsymbol{v}}_{\mathscr{L}}\big(\boldsymbol{x}_0(\boldsymbol{\rho})\big)\Big] \cdot \hat{\boldsymbol{\nu}}_{\mathscr{L}}\big(\boldsymbol{x}_0(\boldsymbol{\rho})\big) \Big\|\boldsymbol{\nu}_{\mathscr{L}}\big(\boldsymbol{x}_0(\boldsymbol{\rho})\big)\Big\| \neq 0, \infty,$$
$$\forall\, \boldsymbol{\rho} \mid \boldsymbol{x}_0(\boldsymbol{\rho}) \in \mathscr{L}, \tag{5.49}$$

provided that the velocity of the boundary is also bounded. The above relationship implies that the BC is never characteristic. Actually, it may be deduced that the determinant of the Jacobian matrix only vanishes if $\boldsymbol{x}_0(\boldsymbol{\rho}) \in \mathscr{N}$.

On the other hand, imposing the IC (5.3), which corresponds to $s_0(\boldsymbol{\rho}) = 0$, $x_0(\boldsymbol{\rho}) = \rho_1$, $y_0(\boldsymbol{\rho}) = \rho_2$, yields

$$\det \mathbf{J}_0\big(\boldsymbol{x}_0(\boldsymbol{\rho}), s_0(\rho_1)\big) = \begin{vmatrix} 1 & 0 & \bar{v}_x\big(y_0(\rho_2)\big) \\ 0 & 1 & \bar{v}_y\big(x_0(\rho_1)\big) \\ 0 & 0 & 1 \end{vmatrix} = 1 \neq 0, \infty. \tag{5.50}$$

References

1. Romano L, Bruzelius F, Jacobson B. Brush tyre models for large camber angles and steering speeds. Vehicle Syst Dyn. Available from: https://doi.org/10.1080/00423114.2020.1854320
2. Romano L, Timpone F, Bruzelius F, Jacobson B. Analytical results in transient brush tyre models: theory for large camber angles and classic solutions with limited friction. Meccanica. Available from: https://doi.org/10.1007/s11012-021-01422-3
3. Evans LC (2010) Partial differential equations, 2nd ed. American Mathematical Society
4. Ockendon JR, Howison S, Lacey A, Movchan A (2003) Applied partial differential equations. Oxford University Press
5. Rugh, WJ (2007) Linear system theory, 2nd ed. Johns Hopkins University
6. Pacejka HB (2012) Tire and vehicle dynamics, 3rd edn. Elsevier/BH, Amsterdam
7. Khalil HK (2002) Nonlinear systems, 3rd edn. Prentice Hall, Upper Saddle River

Chapter 6
Theories of Transient Generation of Tyre Forces and Moment

Abstract The knowledge gained from the analysis of the transient brush theory may be effectively used to derive simplified, pragmatic tyre models to be used in vehicle dynamics simulations. Two different formulations have been proposed in the literature. The first is the so-called single contact point. It consists of a simple class of models, which usually find good agreement with experimental data and are based on a rather intuitive approximation. However, they systematically neglect the dynamics of the bristles inside the contact patch. The second formulation presented in this chapter is the two-regime approach, which also captures the transient effect due to the bristles. It relies on more sophisticated mathematical tools, and, albeit being relatively easy to implement, rarely admit closed-form representations. The two different models are compared together against the exact description introduced in Chap. 4.

The transient models presented in Chaps. 4 and 5 provided many insights on the nonstationary behaviour of the tyre subjected to variable slip inputs. However, these formulations are too complicated to be used for practical applications. Therefore, the aim of the present chapter is to introduce two approximate transient tyre models: the single contact point and the two-regime approach. These describe the tyre dynamics as a system of ODEs, instead of PDEs, and are more appropriate when it comes to vehicle dynamics studies. Despite their simplistic nature, indeed, they are able to capture the salient nonstationary phenomena with good accuracy and may be easily implemented in simulation environments.

In the following, Sect. 6.1 is dedicated to the single contact point models, whilst the two-regime theory is introduced in Sect. 6.2. A comparison between the two formulations is then made in Sect. 6.3. Both the models are derived from the classic transient theory outlined in Chap. 4, and therefore valid for relatively small camber angles.

L. Romano, *Advanced Brush Tyre Modelling*,
SpringerBriefs in Applied Sciences and Technology,
https://doi.org/10.1007/978-3-030-98435-9_6

6.1 Single Contact Point Models

The single contact point models were introduced by Higuchi [1–3] in his excellent dissertation and further extended by Zegelaar and Maurice [4, 5]. The mathematical treatment of the single contact point models presented in this paper is mainly adapted from the formulations by Guiggiani [6], Svendenius [7] and Rill [8, 9].

The overall idea behind the single contact point models is to replace the translational slip variable $\boldsymbol{\sigma}$ with the transient slip $\boldsymbol{\sigma}'$ in the description of the steady-state forces and moment, that is

$$\boldsymbol{F}_t = \boldsymbol{F}_t(\boldsymbol{\sigma}', \varphi), \tag{6.1a}$$

$$M_z = M_z(\boldsymbol{\sigma}', \varphi), \tag{6.1b}$$

or

$$\boldsymbol{F}_t = \boldsymbol{F}_t(\boldsymbol{\kappa}', \varphi), \tag{6.2a}$$

$$M_z = M_z(\boldsymbol{\kappa}', \varphi), \tag{6.2b}$$

or

$$\boldsymbol{F}_t = \boldsymbol{F}_t(\kappa_x', \alpha', \gamma), \tag{6.3a}$$

$$M_z = M_z(\kappa_x', \alpha', \gamma). \tag{6.3b}$$

The steady-state expressions for $\boldsymbol{F}_t$ and M_z may be derived starting from the brush theory, as done in Chap. 3, or better using empirical relationships, like for example Pacejka's Magic Formula [1, 11].

6.1.1 Model Equations

The governing equations of the single contact point models have a slightly different structure depending on the specific choice of slips used to describe the steady-state relationships for the tyre forces and moment.

6.1.1.1 Formulation in Terms of Theoretical Slips

Differentiating Eq. (6.1a) with respect to the time and combining (1.13) with (2.7) yields

$$\dot{\boldsymbol{F}}_t\big(\boldsymbol{\sigma}'(t), \varphi(t)\big) = \nabla_{\boldsymbol{\sigma}'}\boldsymbol{F}_t(\boldsymbol{\sigma}', \varphi)^{\mathrm{T}}\dot{\boldsymbol{\sigma}}'(t) + \frac{\partial \boldsymbol{F}_t(\boldsymbol{\sigma}', \varphi)}{\partial \varphi}\dot{\varphi}(t) = V_{\mathrm{r}}(t)\mathbf{C}'\big(\boldsymbol{\sigma}(t) - \boldsymbol{\sigma}'(t)\big). \tag{6.4}$$

Further simplifications allow to define the concept of the so-called relaxation lengths. In general, the quantities relating to φ are almost irrelevant for the transient generation of tyre forces, and hence may be disregarded [6]. Premultiplying Eq. (6.4) by $\mathbf{S}' = \mathbf{C}'^{-1}$ then gives

$$\mathbf{S}'\tilde{\mathbf{C}}_\sigma\big(\boldsymbol{\sigma}'(t), \varphi(t)\big)\dot{\boldsymbol{\sigma}}'(t) = V_{\mathrm{r}}(t)\big(\boldsymbol{\sigma}(t) - \boldsymbol{\sigma}'(t)\big), \tag{6.5}$$

with the matrix of the generalised tyre stiffnesses defined as in Eq. (1.26a). Equation (6.5) may be recast more conveniently as

$$\tilde{\boldsymbol{\Lambda}}_\sigma\big(\boldsymbol{\sigma}'(t), \varphi(t)\big)\dot{\boldsymbol{\sigma}}'(t) + V_{\mathrm{r}}(t)\boldsymbol{\sigma}'(t) = V_{\mathrm{r}}(t)\boldsymbol{\sigma}(t), \tag{6.6}$$

where the matrix $\tilde{\boldsymbol{\Lambda}}_\sigma(\boldsymbol{\sigma}', \varphi)$ of the generalised theoretical relaxation lengths may be defined as

$$\tilde{\boldsymbol{\Lambda}}_\sigma\big(\boldsymbol{\sigma}', \varphi\big) = \begin{bmatrix} \tilde{\lambda}_{x\sigma_x}\big(\boldsymbol{\sigma}', \varphi\big) & \tilde{\lambda}_{x\sigma_y}\big(\boldsymbol{\sigma}', \varphi\big) \\ \tilde{\lambda}_{y\sigma_x}\big(\boldsymbol{\sigma}', \varphi\big) & \tilde{\lambda}_{y\sigma_y}\big(\boldsymbol{\sigma}', \varphi\big) \end{bmatrix} \triangleq \mathbf{S}'\tilde{\mathbf{C}}_\sigma\big(\boldsymbol{\sigma}', \varphi\big). \tag{6.7}$$

Equation (6.6) is a nonlinear system of two coupled ODEs for the transient theoretical slip vector $\boldsymbol{\sigma}'$, where the conventional slip is interpreted as a time-varying input. The transient theoretical slip $\boldsymbol{\sigma}'$ calculated by means of Eq. (6.6) is then used as input to the corresponding steady-state model for the tyre forces and moment, that is Eqs. (6.1a), (6.1b), treated as static nonlinearities. This variant of the single contact point model is referred to as *full-nonlinear*, since the matrix $\tilde{\boldsymbol{\Lambda}}_\sigma(\boldsymbol{\sigma}', \varphi)$ is itself a function of $\boldsymbol{\sigma}'$.

Example 6.1.1 For a tyre with isotropic tread and rectangular contact patch in pure translational slip conditions ($\boldsymbol{\sigma} \neq 0$, $\varphi = 0$) as in Sect. 3.3, the transient dynamics is described by

$$\tilde{\boldsymbol{\Lambda}}_\sigma\big(\boldsymbol{\sigma}'(t)\big)\dot{\boldsymbol{\sigma}}'(t) + V_{\mathrm{r}}(t)\boldsymbol{\sigma}'(t) = V_{\mathrm{r}}(t)\boldsymbol{\sigma}(t), \tag{6.8a}$$

$$\boldsymbol{F}_t\big(\boldsymbol{\sigma}'(t)\big) = C_\sigma\boldsymbol{\sigma}'(t)\left[1 - \frac{\sigma'(t)}{\sigma^{\mathrm{cr}}} + \frac{1}{3}\left(\frac{\sigma'(t)}{\sigma^{\mathrm{cr}}}\right)^2\right], \tag{6.8b}$$

$$M_z\big(\boldsymbol{\sigma}'(t)\big) = -C_{M\sigma_y}\sigma_y'(t)\left[1 - 3\frac{\sigma'(t)}{\sigma^{\mathrm{cr}}} + 3\left(\frac{\sigma'(t)}{\sigma^{\mathrm{cr}}}\right)^2 - \left(\frac{\sigma'(t)}{\sigma^{\mathrm{cr}}}\right)^3\right], \tag{6.8c}$$

where $\sigma'(t) = \big\|\boldsymbol{\sigma}'(t)\big\|$ and the matrix of generalised theoretical relaxation lengths $\tilde{\boldsymbol{\Lambda}}_\sigma(\boldsymbol{\sigma}')$ reads specifically

$$\tilde{\mathbf{\Lambda}}_\sigma\big(\boldsymbol{\sigma}'(t)\big) = \begin{bmatrix} \dfrac{\tilde{C}_{x\sigma_x}\big(\boldsymbol{\sigma}'(t)\big)}{C'_x} & \dfrac{\tilde{C}_{x\sigma_y}\big(\boldsymbol{\sigma}'(t)\big)}{C'_x} \\ \dfrac{\tilde{C}_{y\sigma_x}\big(\boldsymbol{\sigma}'(t)\big)}{C'_y} & \dfrac{\tilde{C}_{y\sigma_y}\big(\boldsymbol{\sigma}'(t)\big)}{C'_y} \end{bmatrix}, \tag{6.9}$$

with

$$\tilde{C}_{x\sigma_x}\big(\boldsymbol{\sigma}'(t)\big) = \frac{\partial F_x\big(\boldsymbol{\sigma}'(t)\big)}{\partial \sigma'_x} = C_\sigma\left[1 - \frac{2\sigma'_x(t) + \sigma'_y(t)}{\sigma^{\mathrm{cr}}\sigma'(t)} + \left(\frac{\sigma'_x(t)}{\sigma^{\mathrm{cr}}}\right)^2 + \frac{1}{3}\left(\frac{\sigma'_y(t)}{\sigma^{\mathrm{cr}}}\right)^2\right], \tag{6.10a}$$

$$\tilde{C}_{y\sigma_y}\big(\boldsymbol{\sigma}'(t)\big) = \frac{\partial F_y\big(\boldsymbol{\sigma}'(t)\big)}{\partial \sigma'_y} = C_\sigma\left[1 - \frac{2\sigma'_y(t) + \sigma'_x(t)}{\sigma^{\mathrm{cr}}\sigma'(t)} + \left(\frac{\sigma'_y(t)}{\sigma^{\mathrm{cr}}}\right)^2 + \frac{1}{3}\left(\frac{\sigma'_x(t)}{\sigma^{\mathrm{cr}}}\right)^2\right], \tag{6.10b}$$

$$\tilde{C}_{x\sigma_y}\big(\boldsymbol{\sigma}'(t)\big) \equiv \tilde{C}_{y\sigma_x}\big(\boldsymbol{\sigma}'(t)\big) = \frac{\partial F_x\big(\boldsymbol{\sigma}'(t)\big)}{\partial \sigma'_y} = \frac{\partial F_y\big(\boldsymbol{\sigma}'(t)\big)}{\partial \sigma'_x} = -C_\sigma \frac{\sigma'_x(t)\sigma'_y(t)}{\sigma^{\mathrm{cr}}}\left[\frac{1}{\sigma'(t)} - \frac{2}{3\sigma^{\mathrm{cr}}}\right]. \tag{6.10c}$$

In practical applications, the matrix of theoretical relaxation lengths is often assumed to be constant and replaced with the conventional one, reading

$$\mathbf{\Lambda}_\sigma = \begin{bmatrix} \lambda_{x\sigma_x} & \lambda_{x\sigma_y} \\ \lambda_{y\sigma_x} & \lambda_{y\sigma_y} \end{bmatrix} \triangleq \tilde{\mathbf{\Lambda}}_\sigma(\mathbf{0}, 0) = \mathbf{S}'\tilde{\mathbf{C}}_\sigma(\mathbf{0}, 0) = \mathbf{S}'\mathbf{C}_\sigma. \tag{6.11}$$

It may be deduced from Eq. (6.11) that considering the conventional relaxation matrix $\mathbf{\Lambda}_\sigma$ in place of the generalised one is equivalent to approximate the generalised theoretical stiffness matrix with the conventional one in Eq. (6.5). This model variation is often referred to as *semi-nonlinear single contact point*, since in this case the system (6.6) is linear, but the tyre forces are still modelled as static nonlinearities.

Example 6.1.2 When the matrix of the generalised theoretical relaxation lengths is approximated by the conventional one, Eq. (6.8a) becomes

$$\mathbf{\Lambda}_\sigma\dot{\boldsymbol{\sigma}}'(t) + V_{\mathrm{r}}(t)\boldsymbol{\sigma}'(t) = V_{\mathrm{r}}(t)\boldsymbol{\sigma}(t), \tag{6.12a}$$

with

$$\mathbf{\Lambda}_\sigma = \begin{bmatrix} \lambda_{\sigma_x} & 0 \\ 0 & \lambda_{\sigma_y} \end{bmatrix} = \begin{bmatrix} \dfrac{C_\sigma}{C'_x} & 0 \\ 0 & \dfrac{C_\sigma}{C'_y} \end{bmatrix}, \tag{6.13}$$

where the relaxation lengths have been renamed $\lambda_{x\sigma_x} = \lambda_{\sigma_x}$ and $\lambda_{y\sigma_y} = \lambda_{\sigma_y}$ without ambiguity. The transient tyre forces and moment are still described by Eqs. (6.8b) and (6.8c).

When the relaxation matrix is constant and the expressions for the tyre forces and moment are also linear, the model is commonly referred to as *linear single contact point.*

Example 6.1.3 The dynamics of the transient slip $\boldsymbol{\sigma}'$ is governed by Eq. (6.12a), with $\boldsymbol{\Lambda}_\sigma$ as in Eq. (6.13), and the steady-state relationships for the tyre forces and moment are modelled as follows:

$$F_t\big(\boldsymbol{\sigma}'(t)\big) = C_\sigma \boldsymbol{\sigma}'(t), \tag{6.14a}$$

$$M_z\big(\boldsymbol{\sigma}'(t)\big) = -C_{M\sigma_y}\sigma_y'(t). \tag{6.14b}$$

6.1.1.2 Formulation in Terms of Practical Slips

Differentiating Eq. (6.2a) with respect to the time and combining (1.14) with (2.7) yields

$$\dot{\boldsymbol{F}}_t\big(\boldsymbol{\kappa}'(t), \varphi(t)\big) = \nabla_{\boldsymbol{\kappa}'}\boldsymbol{F}_t\big(\boldsymbol{\kappa}', \varphi\big)^{\mathrm{T}}\dot{\boldsymbol{\kappa}}'(t) + \frac{\partial \boldsymbol{F}_t\big(\boldsymbol{\kappa}', \varphi\big)}{\partial \varphi}\dot{\varphi}(t) = V_{Cx}(t)\mathbf{C}'\big(\boldsymbol{\kappa}(t) - \boldsymbol{\kappa}'(t)\big). \tag{6.15}$$

Similarly to what done before, Eq. (6.15) may be restated as

$$\tilde{\boldsymbol{\Lambda}}_\kappa\big(\boldsymbol{\kappa}'(t), \varphi(t)\big)\dot{\boldsymbol{\kappa}}'(t) + V_{Cx}(t)\boldsymbol{\kappa}'(t) = V_{Cx}(t)\boldsymbol{\kappa}(t), \tag{6.16}$$

with the matrix $\tilde{\boldsymbol{\Lambda}}_\kappa(\boldsymbol{\kappa}', \varphi)$ of the generalised practical relaxation lengths reading

$$\tilde{\boldsymbol{\Lambda}}_\kappa\big(\boldsymbol{\kappa}', \varphi\big) = \begin{bmatrix} \tilde{\lambda}_{x\kappa_x}\big(\boldsymbol{\kappa}', \varphi\big) & \tilde{\lambda}_{x\kappa_y}\big(\boldsymbol{\kappa}', \varphi\big) \\ \tilde{\lambda}_{y\kappa_x}\big(\boldsymbol{\kappa}', \varphi\big) & \tilde{\lambda}_{y\kappa_y}\big(\boldsymbol{\kappa}', \varphi\big) \end{bmatrix} \triangleq \mathbf{S}'\tilde{\mathbf{C}}_\kappa\big(\boldsymbol{\kappa}', \varphi\big). \tag{6.17}$$

In this case, Eqs. (6.16) is a nonlinear system of two coupled ODEs for the transient practical slip vector $\boldsymbol{\kappa}'$, which represents the input to Eqs. (6.2a), (6.2b).

Starting from Eq. (6.17), the matrix of conventional practical relaxation lengths may be defined as

$$\boldsymbol{\Lambda}_\kappa = \begin{bmatrix} \lambda_{x\kappa_x} & \lambda_{x\kappa_y} \\ \lambda_{y\kappa_x} & \lambda_{y\kappa_y} \end{bmatrix} \triangleq \tilde{\boldsymbol{\Lambda}}_\kappa(\mathbf{0}, 0) = \mathbf{S}'\tilde{\mathbf{C}}_\kappa(\mathbf{0}, 0) = \mathbf{S}'\mathbf{C}_\kappa. \tag{6.18}$$

6.1.1.3 Formulation in Terms of Geometrical Slips

Differentiating Eq. (6.3a) with respect to the time and combining (1.14) and (1.15) with (2.7) yields

$$
\dot{\boldsymbol{F}}_t\big(\kappa_x'(t), \alpha'(t), \gamma(t)\big) = \begin{bmatrix} \dfrac{\partial F_x\big(\kappa_x', \alpha', \gamma\big)}{\partial \kappa_x'} & \dfrac{\partial F_x\big(\kappa_x', \alpha', \gamma\big)}{\partial \alpha'} \\ \dfrac{\partial F_y\big(\kappa_x', \alpha', \gamma\big)}{\partial \kappa_x'} & \dfrac{\partial F_y\big(\kappa_x', \alpha', \gamma\big)}{\partial \alpha'} \end{bmatrix} \begin{bmatrix} \dot{\kappa}_x'(t) \\ \dot{\alpha}'(t) \end{bmatrix}
$$

$$
+ \frac{\partial \boldsymbol{F}_t\big(\kappa_x', \alpha'(t), \gamma\big)}{\partial \gamma} \dot{\gamma}(t) = V_{Cx}(t)\mathbf{C}' \begin{bmatrix} \kappa_x'(t) - \kappa_x(t) \\ \tan \alpha'(t) - \tan \alpha(t) \end{bmatrix}. \tag{6.19}
$$

Eq. (6.5) may be recast more conveniently as

$$
\begin{bmatrix} \tilde{\lambda}_{x\kappa_x}\big(\kappa_x'(t), \alpha'(t), \gamma(t)\big) & \tilde{\lambda}_{x\alpha}\big(\kappa_x'(t), \alpha'(t), \gamma(t)\big) \\ \tilde{\lambda}_{y\kappa_x}\big(\kappa_x'(t), \alpha'(t), \gamma(t)\big) & \tilde{\lambda}_{y\alpha}\big(\kappa_x'(t), \alpha'(t), \gamma(t)\big) \end{bmatrix} \begin{bmatrix} \dot{\kappa}_x'(t) \\ \dot{\alpha}'(t) \end{bmatrix} + V_{Cx}(t) \begin{bmatrix} \kappa_x'(t) \\ \alpha'(t) \end{bmatrix} = V_{Cx}(t) \begin{bmatrix} \kappa_x(t) \\ \alpha(t) \end{bmatrix}, \tag{6.20}
$$

with

$$
\begin{bmatrix} \tilde{\lambda}_{x\alpha}\big(\kappa_x'(t), \alpha'(t), \gamma(t)\big) \\ \tilde{\lambda}_{y\alpha}\big(\kappa_x'(t), \alpha'(t), \gamma(t)\big) \end{bmatrix} = \mathbf{S}' \begin{bmatrix} \tilde{C}_{x\alpha}\big(\kappa_x, \alpha, \gamma, \varphi_\psi\big) \\ \tilde{C}_{y\alpha}\big(\kappa_x, \alpha, \gamma, \varphi_\psi\big) \end{bmatrix}. \tag{6.21}
$$

In this case, Eqs. (6.20) is a nonlinear system of two coupled ODEs for the transient geometrical slips κ_x' and α', which represent the input to Eqs. (6.3a), (6.3b).

Starting from Eq. (6.21), the matrix of conventional geometrical relaxation lengths may be defined as

$$
\begin{bmatrix} \lambda_{x\alpha} \\ \lambda_{y\alpha} \end{bmatrix} \triangleq \begin{bmatrix} \tilde{\lambda}_{x\alpha}(0, 0, 0) \\ \tilde{\lambda}_{y\alpha}(0, 0, 0) \end{bmatrix} = \mathbf{S}' \begin{bmatrix} \tilde{C}_{x\alpha}(0, 0, 0) \\ \tilde{C}_{y\alpha}(0, 0, 0) \end{bmatrix} = \mathbf{S}' \begin{bmatrix} C_{x\alpha} \\ C_{y\alpha} \end{bmatrix}. \tag{6.22}
$$

6.1.2 Relationships Between Relaxation Matrices

Similar to what done in Subsect. 1.4.1 for the stiffness matrices, the relationships between the generalised relaxation lengths may be studied with the aid of the transformations given by Eqs. (1.9) and (1.10). More specifically, between the theoretical and practical generalised relaxation matrices, the following relationships may be found:

$$\tilde{\mathbf{\Lambda}}_{\kappa}\big(\boldsymbol{\kappa}', \varphi\big) = \tilde{\mathbf{\Lambda}}_{\sigma}\big(\boldsymbol{\sigma}', \varphi\big)\bigg|_{\boldsymbol{\sigma}=\frac{\boldsymbol{\kappa}'}{1+\kappa_x'}} \begin{bmatrix} \dfrac{1}{\big(1+\kappa_x'\big)^2} & 0 \\ -\dfrac{\kappa_y'}{\big(1+\kappa_x'\big)^2} & \dfrac{1}{1+\kappa_x'} \end{bmatrix}, \tag{6.23a}$$

$$\tilde{\mathbf{\Lambda}}_{\sigma}\big(\boldsymbol{\sigma}', \varphi\big) = \tilde{\mathbf{\Lambda}}_{\kappa}\big(\boldsymbol{\kappa}', \varphi\big)\bigg|_{\boldsymbol{\kappa}'=\frac{\boldsymbol{\sigma}'}{1-\sigma_x'}} \begin{bmatrix} \dfrac{1}{\big(1-\sigma_x'\big)^2} & 0 \\ -\dfrac{\sigma_y'}{\big(1-\sigma_x'\big)^2} & \dfrac{1}{1-\sigma_x'} \end{bmatrix}. \tag{6.23b}$$

Clearly it holds that

$$\mathbf{\Lambda}_{\kappa} \triangleq \tilde{\mathbf{\Lambda}}_{\kappa}(\mathbf{0}, 0) = \tilde{\mathbf{\Lambda}}_{\sigma}(\mathbf{0}, 0) \triangleq \mathbf{\Lambda}_{\sigma}. \tag{6.24}$$

On the other hand, the practical and geometrical generalised relaxation lengths are related by the transformations

$$\begin{bmatrix} \tilde{\lambda}_{x\alpha}\big(\kappa_x', \alpha', \gamma\big) \\ \tilde{\lambda}_{y\alpha}\big(\kappa_x', \alpha', \gamma\big) \end{bmatrix} = \frac{1}{\cos^2\big(\alpha'\big)} \begin{bmatrix} \tilde{\lambda}_{x\kappa_y}\big(\boldsymbol{\kappa}', \varphi\big) \\ \tilde{\lambda}_{y\kappa_y}\big(\boldsymbol{\kappa}', \varphi\big) \end{bmatrix}\Bigg|_{\kappa_y'=\tan\alpha', \varphi=(1-\varepsilon_\gamma)\frac{\sin\gamma}{R_r}}, \tag{6.25a}$$

$$\begin{bmatrix} \tilde{\lambda}_{x\kappa_y}\big(\boldsymbol{\kappa}', \varphi\big) \\ \tilde{\lambda}_{y\kappa_y}\big(\boldsymbol{\kappa}', \varphi\big) \end{bmatrix} = \frac{1}{1+\kappa_y'} \begin{bmatrix} \tilde{\lambda}_{x\alpha}\big(\kappa_x', \alpha', \gamma\big) \\ \tilde{\lambda}_{y\alpha}\big(\kappa_x', \alpha', \gamma\big) \end{bmatrix}\Bigg|_{\alpha'=\arctan\kappa_y', \gamma=\arcsin\left(\frac{\varphi R_r}{1-\varepsilon_\gamma}\right)}, \tag{6.25b}$$

and consequently

$$\begin{bmatrix} \lambda_{x\alpha} \\ \lambda_{y\alpha} \end{bmatrix} \triangleq \begin{bmatrix} \tilde{\lambda}_{x\alpha}(0, 0, 0) \\ \tilde{\lambda}_{y\alpha}(0, 0, 0) \end{bmatrix} = \begin{bmatrix} \tilde{\lambda}_{x\kappa_y}(\mathbf{0}, 0) \\ \tilde{\lambda}_{y\kappa_y}(\mathbf{0}, 0) \end{bmatrix} \triangleq \begin{bmatrix} \lambda_{x\kappa_y} \\ \lambda_{y\kappa_y} \end{bmatrix}. \tag{6.26}$$

Up to this point, it should be acknowledged that the single contact point models well approximate the real dynamics of a tyre in transient slip conditions. This also legitimates their ubiquitous presence in vehicle dynamics [10]. However, they systematically neglect transient effects due to the deflection of the bristles in the contact patch. As already discussed in Chaps. 4 and 5, this constitutes an essential part of the transient process for the generation of the tyre forces and moments. Moreover, using a similar rationale as in [16], it may be shown that

$$\dot{\boldsymbol{F}}_t(t) = \iint_{\mathscr{P}} \frac{\partial \boldsymbol{q}_t(\boldsymbol{x}, t)}{\partial t} \,\mathrm{d}\boldsymbol{x}, \tag{6.27a}$$

$$\dot{M}_z(t) = \iint_{\mathscr{P}} x\frac{\partial q_y(\boldsymbol{x}, t)}{\partial t} - y\frac{\partial q_x(\boldsymbol{x}, t)}{\partial t} \,\mathrm{d}\boldsymbol{x}, \tag{6.27b}$$

and therefore, if the bristles are in steady-state conditions, the tyre forces and moment are also constant.

6.2 Two-Regime Theory

The two-regime theory is based directly on the structure of the governing PDEs of the classic brush theory [12], and describes the tyre dynamics as a functional interpolation between two opposite behaviours occurring at high and low rolling speeds. Specifically, the two-regime models may be derived by considering two approximate sets of relationships existing between the slip variables and the tyre forces and moment. To better understand this concept, it may be useful to recast Eq. (2.15) in the time domain:

$$\boldsymbol{v}_{\mathrm{s}}(\boldsymbol{x},t) = -V_{\mathrm{r}}(t)\boldsymbol{\sigma}'(t) - V_{\mathrm{r}}(t)\mathbf{A}_{\varphi}(t)\boldsymbol{x} + \frac{\partial \boldsymbol{u}_t(\boldsymbol{x},t)}{\partial t} - V_{\mathrm{r}}(t)\frac{\partial \boldsymbol{u}_t(\boldsymbol{x},t)}{\partial x}, \quad (\boldsymbol{x},t) \in \mathring{\mathscr{P}} \times \mathbb{R}_{>0}, \tag{6.28}$$

with $\boldsymbol{v}_{\mathrm{s}}(\boldsymbol{x},t) = V_{\mathrm{r}}(t)\bar{\boldsymbol{v}}_{\mathrm{s}}(\boldsymbol{x},t)$. It may be anticipated from Eq. (6.28) that, for $V_{\mathrm{r}}(t) \to 0$, the first partial derivative will be preponderant on the second one; *vice versa*, for $V_{\mathrm{r}} \to \infty$, the time derivative will be negligible. Therefore, at low speeds, approximate relationships are sought by neglecting the partial derivative with respect to the space variable x in Eq. (6.28), and may be cast in the general form

$$\boldsymbol{\sigma} \approx \frac{1}{V_{\mathrm{r}}}\breve{\boldsymbol{\sigma}}\Big(\dot{\boldsymbol{F}}_t, \dot{M}_z\Big), \tag{6.29a}$$

$$\varphi \approx \frac{1}{V_{\mathrm{r}}}\breve{\varphi}\Big(\dot{\boldsymbol{F}}_t, \dot{M}_z\Big), \tag{6.29b}$$

and

$$\boldsymbol{\kappa} \approx \frac{1}{V_{Cx}}\breve{\boldsymbol{\kappa}}\Big(\dot{\boldsymbol{F}}_t, \dot{M}_z\Big), \tag{6.30a}$$

$$\varphi \approx \frac{1}{V_{\mathrm{r}}}\breve{\varphi}\Big(\dot{\boldsymbol{F}}_t, \dot{M}_z\Big), \tag{6.30b}$$

and

$$\kappa_x \approx \frac{1}{V_{Cx}}\breve{\kappa}_x\Big(\dot{\boldsymbol{F}}_t, \dot{M}_z\Big), \tag{6.31a}$$

$$\tan\alpha \approx \frac{1}{V_{Cx}}\tan\breve{\alpha}\Big(\dot{\boldsymbol{F}}_t, \dot{M}_z\Big), \tag{6.31b}$$

$$\sin\gamma \approx \frac{1}{V_{\mathrm{r}}}\sin\breve{\gamma}\Big(\dot{\boldsymbol{F}}_t, \dot{M}_z\Big). \tag{6.31c}$$

The functions on the right-hand side of Eqs. (6.29a), (6.29b), (6.30a), (6.30b) and (6.31a), (6.31b), (6.31c) are called *theoretical*, *practical* and *geometrical sliding functions*, respectively, since they provide a set of relationships between the derivative of the tangential forces and moment and the sliding velocities.

On the other hand, at high speeds (rolling V_r or longitudinal V_{Cx}), the corresponding relationships are postulated as

$$\boldsymbol{\sigma} \approx \hat{\boldsymbol{\sigma}}(\boldsymbol{F}_t, M_z), \tag{6.32a}$$
$$\varphi \approx \hat{\varphi}(\boldsymbol{F}_t, M_z), \tag{6.32b}$$

and

$$\boldsymbol{\kappa} \approx \hat{\boldsymbol{\kappa}}(\boldsymbol{F}_t, M_z), \tag{6.33a}$$
$$\varphi \approx \hat{\varphi}(\boldsymbol{F}_t, M_z), \tag{6.33b}$$

and

$$\kappa_x \approx \hat{\kappa}_x(\boldsymbol{F}_t, M_z), \tag{6.34a}$$
$$\tan\alpha \approx \tan\hat{\alpha}(\boldsymbol{F}_t, M_z), \tag{6.34b}$$
$$\sin\gamma \approx \sin\hat{\gamma}(\boldsymbol{F}_t, M_z), \tag{6.34c}$$

and are obtained by disregarding the partial derivatives with respect to the time in Eq. (6.28). Therefore, the resulting expressions coincide with the steady-state slip-tyre relationships already introduced in Subsect. 1.4.3. More precisely, it should be noticed that the relationships for $\tan\hat{\alpha}(\cdot,\cdot)$ and $\sin\hat{\gamma}(\cdot,\cdot)$ in Eqs. (6.34a), (6.34b), (6.34c) must be derived by the composition of Eqs. (1.47b) and (1.47c) with the $\tan(\cdot)$ and $\sin(\cdot)$ functions, respectively. Functions of the type (6.32a), (6.32b), (6.33a), (6.33b) and (6.34a), (6.34b), (6.34c) may be always obtained owing to the invertibility conditions stated by Eqs. (1.48) and are generally referred to as *slip functions* (*theoretical*, *practical* and *geometrical*, respectively).

Interpolating between the sliding and slip functions using the rolling and longitudinal speed as weighting factors yields

$$\boldsymbol{\sigma} = \frac{1}{V_r}\breve{\boldsymbol{\sigma}}\left(\dot{\boldsymbol{F}}_t, \dot{M}_z\right) + \hat{\boldsymbol{\sigma}}(\boldsymbol{F}_t, M_z), \tag{6.35a}$$
$$\varphi = \frac{1}{V_r}\breve{\varphi}\left(\dot{\boldsymbol{F}}_t, \dot{M}_z\right) + \hat{\varphi}(\boldsymbol{F}_t, M_z), \tag{6.35b}$$

and

$$\boldsymbol{\kappa} = \frac{1}{V_{Cx}}\breve{\boldsymbol{\kappa}}\left(\dot{\boldsymbol{F}}_t, \dot{M}_z\right) + \hat{\boldsymbol{\kappa}}(\boldsymbol{F}_t, M_z), \tag{6.36a}$$
$$\varphi = \frac{1}{V_r}\breve{\varphi}\left(\dot{\boldsymbol{F}}_t, \dot{M}_z\right) + \hat{\varphi}(\boldsymbol{F}_t, M_z), \tag{6.36b}$$

and

$$\kappa_x = \frac{1}{V_{Cx}}\check{\kappa}_x\Big(\dot{\boldsymbol{F}}_t, \dot{M}_z\Big) + \hat{\kappa}_x(\boldsymbol{F}_t, M_z), \tag{6.37a}$$

$$\tan\alpha = \frac{1}{V_{Cx}}\tan\check{\alpha}\Big(\dot{\boldsymbol{F}}_t, \dot{M}_z\Big) + \tan\hat{\alpha}(\boldsymbol{F}_t, M_z), \tag{6.37b}$$

$$\sin\gamma = \frac{1}{V_r}\sin\check{\gamma}\Big(\dot{\boldsymbol{F}}_t, \dot{M}_z\Big) + \sin\hat{\gamma}(\boldsymbol{F}_t, M_z), \tag{6.37c}$$

which represent the general formulations for the two-regime transient tyre model. A quick inspection reveals that if the sliding functions have an isolated equilibrium in the origin, then the equilibria of the dynamical systems described by Eqs. (6.35a), (6.35b), (6.36a), (6.36b) and (6.37a), (6.37b), (6.37c) coincide with the solutions of the steady-state tyre-slip relationships. Actually, the sliding functions can be constructed to satisfy this requirement, as shown in the next Subsect. 6.2.1. Furthermore, the two-regime models usually outperform the single contact point formulation in predicting the transient tyre forces and moment. The main limitation is that the steady-state tyre-slip relationships are often only locally diffeomorphic, and therefore the sliding functions in Eqs. (6.29a), (6.29b), (6.30a), (6.30b) and (6.31a), (6.31b), (6.31c) are restricted to a limited domain.

6.2.1 Derivation of the Sliding Functions

The sliding functions have a different structure depending on the specific choice of slip variables.

6.2.1.1 Theoretical Sliding Functions

The theoretical sliding functions appearing in Eqs. (6.29a), (6.29b) may be derived by substitution of Eq. (6.28) into Eqs. (6.27a), (6.27b). Disregarding the partial derivative with respect to the longitudinal coordinate and integrating under vanishing sliding assumptions, that is over the entire contact patch, yield the linear system

$$\begin{bmatrix} A_{\mathscr{P}}\mathbf{K}_t & \mathbf{K}_t\tilde{\mathbf{I}}\boldsymbol{S}_{\mathscr{P}} \\ \mathbf{I}_\sigma & I_\varphi \end{bmatrix}\begin{bmatrix} \boldsymbol{\sigma} \\ \varphi \end{bmatrix} \approx \frac{1}{V_r}\begin{bmatrix} \mathbf{I} + A_{\mathscr{P}}\mathbf{K}_t\mathbf{S}' & \mathbf{0} \\ \mathbf{I}_\sigma\mathbf{S}' & 1 \end{bmatrix}\begin{bmatrix} \dot{\boldsymbol{F}}_t \\ \dot{M}_z \end{bmatrix}. \tag{6.38}$$

with

$$A_{\mathscr{P}} \triangleq \iint_{\mathscr{P}} \mathrm{d}\boldsymbol{x}, \tag{6.39a}$$

$$\boldsymbol{S}_{\mathscr{P}} \triangleq \iint_{\mathscr{P}} \boldsymbol{x}\,\mathrm{d}\boldsymbol{x}, \tag{6.39b}$$

$$\mathbf{I}_{\sigma} = \begin{bmatrix} I_{\sigma_x} & I_{\sigma_y} \end{bmatrix} \triangleq \begin{bmatrix} \iint_{\mathscr{P}} k_{yx}x - k_{xx}y \, \mathrm{d}\boldsymbol{x} & \iint_{\mathscr{P}} k_{yy}x - k_{xy}y \, \mathrm{d}\boldsymbol{x} \end{bmatrix}, \tag{6.39c}$$

$$I_{\varphi} \triangleq \iint_{\mathscr{P}} k_{yy}x^2 - (k_{xy} + k_{yx})xy + k_{xx}y^2 \, \mathrm{d}\boldsymbol{x}, \tag{6.39d}$$

and the matrix $\tilde{\mathbf{I}}$ defined as in Eq. (4.48b).

Solving Eq. (6.38) for the theoretical slip and spin variables yields

$$\begin{bmatrix} \boldsymbol{\sigma} \\ \varphi \end{bmatrix} \approx \frac{1}{V_{\mathrm{r}}} \begin{bmatrix} \check{\boldsymbol{\sigma}}\left(\dot{\boldsymbol{F}}_t, \dot{M}_z\right) \\ \check{\varphi}\left(\dot{\boldsymbol{F}}_t, \dot{M}_z\right) \end{bmatrix} \triangleq \frac{1}{V_{\mathrm{r}}} \begin{bmatrix} \mathbf{S}'_F & \mathbf{S}'_M \\ \mathbf{S}'_{\varphi F} & S'_{\varphi M} \end{bmatrix} \begin{bmatrix} \dot{\boldsymbol{F}}_t \\ \dot{M}_z \end{bmatrix}, \tag{6.40}$$

with the matrix of generalised compliances defined as

$$\begin{bmatrix} \mathbf{S}'_F & \mathbf{S}'_M \\ \mathbf{S}'_{\varphi F} & S'_{\varphi M} \end{bmatrix} \triangleq \begin{bmatrix} A_{\mathscr{P}}\mathbf{K}_t & \mathbf{K}_t\tilde{\mathbf{I}}S_{\mathscr{P}} \\ \mathbf{I}_{\sigma} & I_{\varphi} \end{bmatrix}^{-1} \begin{bmatrix} \mathbf{I} + A_{\mathscr{P}}\mathbf{K}_t\mathbf{S}' & \mathbf{0} \\ \mathbf{I}_{\sigma}\mathbf{S}' & 1 \end{bmatrix}. \tag{6.41}$$

The functions $\check{\boldsymbol{\sigma}}(\cdot, \cdot)$ and $\check{\varphi}(\cdot, \cdot)$ appearing in Eq. (6.40) are the sought theoretical sliding functions.

6.2.1.2 Practical Sliding Functions

Although they could be deduced starting from the integral equations for the derivatives of the tyre forces and moment, it is perhaps more convenient to derive the theoretical sliding functions directly from Eq. (6.40) with the aid of the transformations (1.10)

$$\boldsymbol{\kappa} \approx \frac{1 + \kappa_x}{V_{\mathrm{r}}} \check{\boldsymbol{\sigma}}\left(\dot{\boldsymbol{F}}_t, \dot{M}_z\right) = \frac{1}{V_{Cx}} \check{\boldsymbol{\sigma}}\left(\dot{\boldsymbol{F}}_t, \dot{M}_z\right) \triangleq \frac{1}{V_{Cx}} \check{\boldsymbol{\kappa}}\left(\dot{\boldsymbol{F}}_t, \dot{M}_z\right). \tag{6.42}$$

It may be easily inferred from Eq. (6.43) that the longitudinal and lateral theoretical and practical sliding functions are formally identical, that is $\check{\boldsymbol{\kappa}}(\cdot, \cdot) = \check{\boldsymbol{\sigma}}(\cdot, \cdot)$. Therefore, at low speed, it may be written as

$$\begin{bmatrix} \boldsymbol{\kappa} \\ \varphi \end{bmatrix} \approx \begin{bmatrix} \frac{1}{V_{Cx}} \check{\boldsymbol{\kappa}}\left(\dot{\boldsymbol{F}}_t, \dot{M}_z\right) \\ \frac{1}{V_{\mathrm{r}}} \check{\varphi}\left(\dot{\boldsymbol{F}}_t, \dot{M}_z\right) \end{bmatrix} \triangleq \begin{bmatrix} \frac{1}{V_{Cx}} \mathbf{S}'_F & \frac{1}{V_{Cx}} \mathbf{S}'_M \\ \frac{1}{V_{\mathrm{r}}} \mathbf{S}'_{\varphi F} & \frac{1}{V_{\mathrm{r}}} S'_{\varphi M} \end{bmatrix} \begin{bmatrix} \dot{\boldsymbol{F}}_t \\ \dot{M}_z \end{bmatrix}. \tag{6.43}$$

6.2.1.3 Geometrical Sliding Functions

Neglecting the contribution of the turn spin to φ, the geometrical sliding functions may be derived combining Eqs. (6.43) with (1.6) and (1.11)

$$\tan\alpha \approx \frac{1}{V_{Cx}}\breve{\kappa}_y\left(\dot{\boldsymbol{F}}_t, \dot{M}_z\right) \triangleq \frac{1}{V_{Cx}} \tan\breve{\alpha}\left(\dot{\boldsymbol{F}}_t, \dot{M}_z\right), \tag{6.44a}$$

$$\sin\gamma \approx \frac{R_{\mathrm{r}}}{(1-\varepsilon_\gamma)V_{\mathrm{r}}}\breve{\varphi}\left(\dot{\boldsymbol{F}}_t, \dot{M}_z\right) \triangleq \frac{1}{V_{\mathrm{r}}} \sin\breve{\gamma}\left(\dot{\boldsymbol{F}}_t, \dot{M}_z\right), \tag{6.44b}$$

which implies that $\tan\breve{\alpha}(\cdot,\cdot) = \breve{\kappa}_y(\cdot,\cdot)$ and

$$\sin\breve{\gamma}\left(\dot{\boldsymbol{F}}_t, \dot{M}_z\right) = \frac{R_{\mathrm{r}}}{1-\varepsilon_\gamma}\breve{\varphi}\left(\dot{\boldsymbol{F}}_t, \dot{M}_z\right) = \frac{R_{\mathrm{r}}}{1-\varepsilon_\gamma}\begin{bmatrix}\mathbf{S}'_{\varphi F} & S'_{\varphi M}\end{bmatrix}\begin{bmatrix}\dot{\boldsymbol{F}}_t \\ \dot{M}_z\end{bmatrix} \triangleq \begin{bmatrix}\mathbf{S}'_{\gamma F} & S'_{\gamma M}\end{bmatrix}\begin{bmatrix}\dot{\boldsymbol{F}}_t \\ \dot{M}_z\end{bmatrix}. \tag{6.45}$$

Therefore

$$\begin{bmatrix}\kappa_x \\ \tan\alpha \\ \varphi\end{bmatrix} \approx \begin{bmatrix}\frac{1}{V_{Cx}}\breve{\kappa}_x\left(\dot{\boldsymbol{F}}_t, \dot{M}_z\right) \\ \frac{1}{V_{Cx}}\tan\breve{\alpha}\left(\dot{\boldsymbol{F}}_t, \dot{M}_z\right) \\ \frac{1}{V_{\mathrm{r}}}\breve{\varphi}\left(\dot{\boldsymbol{F}}_t, \dot{M}_z\right)\end{bmatrix} \triangleq \begin{bmatrix}\frac{1}{V_{Cx}}S'_{xF_x} & \frac{1}{V_{Cx}}S'_{xF_y} & \frac{1}{V_{Cx}}S'_{xM} \\ \frac{1}{V_{Cx}}S'_{yF_x} & \frac{1}{V_{Cx}}S'_{yF_y} & \frac{1}{V_{Cx}}S'_{yM} \\ \frac{1}{V_{\mathrm{r}}}S'_{\gamma F_x} & \frac{1}{V_{\mathrm{r}}}S'_{\gamma F_y} & \frac{1}{V_{\mathrm{r}}}S'_{\gamma M}\end{bmatrix}\begin{bmatrix}\dot{\boldsymbol{F}}_t \\ \dot{M}_z\end{bmatrix}. \tag{6.46}$$

All the sliding functions obtained with the proposed method are linear in the derivative of the tyre forces and moment, and therefore only vanish at the origin, as briefly mentioned before.

6.2.2 *Model Equations*

The governing equations of the two-regime models have a slightly different structure depending on the specific choice of slip variables.

6.2.2.1 Formulation in Terms of Theoretical Slips

Combining Eqs. (6.32a), (6.32b) and (6.40) according to Eqs. (6.35a), (6.35b) and inverting yields

$$\begin{bmatrix}\dot{\boldsymbol{F}}_t(t) \\ \dot{M}_z(t)\end{bmatrix} = V_{\mathrm{r}}(t)\begin{bmatrix}\mathbf{C}'_\sigma & \mathbf{C}'_\varphi \\ \mathbf{C}'_{M\sigma} & C'_{M\varphi}\end{bmatrix}\begin{bmatrix}\boldsymbol{\sigma}(t) - \hat{\boldsymbol{\sigma}}\left(\boldsymbol{F}_t(t), M_z(t)\right) \\ \varphi(t) - \hat{\varphi}\left(\boldsymbol{F}_t(t), M_z(t)\right)\end{bmatrix}, \tag{6.47}$$

where the matrix of the enhanced theoretical slip and spin stiffnesses is defined as

$$\begin{bmatrix} \mathbf{C}'_\sigma & \mathbf{C}'_\varphi \\ \mathbf{C}'_{M\sigma} & C'_{M\varphi} \end{bmatrix} \triangleq \begin{bmatrix} \mathbf{S}'_F & \mathbf{S}'_M \\ \mathbf{S}'_{\varphi F} & S'_{\varphi M} \end{bmatrix}^{-1}. \tag{6.48}$$

Equation (6.47) represents a system of three coupled ODEs for the time-varying tyre forces and moments. Opposed to the single contact point model, Eq. (6.47) does not require the calculation of the transient slip to yield an analytical expression for $\boldsymbol{F}_t(t)$ and $M_z(t)$. Actually, the notion of transient slip is completely absent in the two-regime formulation. However, it is still possible to provide an interpretation in terms of relaxation length by recasting Eq. (6.47) as

$$\begin{bmatrix} \boldsymbol{\Lambda}'_\sigma & \boldsymbol{\Lambda}'_\varphi \\ \boldsymbol{\Lambda}'_{M\sigma} & \lambda'_{M\varphi} \end{bmatrix} \begin{bmatrix} \dot{\boldsymbol{F}}_t(t) \\ \dot{M}_z(t) \end{bmatrix} = V_{\mathrm{r}}(t) \begin{bmatrix} \mathbf{C}_\sigma & \mathbf{C}_\varphi \\ \mathbf{C}_{M\sigma} & C_{M\varphi} \end{bmatrix} \begin{bmatrix} \boldsymbol{\sigma}(t) - \hat{\boldsymbol{\sigma}}\big(\boldsymbol{F}_t(t), M_z(t)\big) \\ \varphi(t) - \hat{\varphi}\big(\boldsymbol{F}_t(t), M_z(t)\big) \end{bmatrix}, \tag{6.49}$$

where the matrix of enhanced theoretical relaxation lengths is given by

$$\begin{bmatrix} \boldsymbol{\Lambda}'_\sigma & \boldsymbol{\Lambda}'_\varphi \\ \boldsymbol{\Lambda}'_{M\sigma} & \lambda'_{M\varphi} \end{bmatrix} = \begin{bmatrix} \lambda'_{x\sigma_x} & \lambda'_{x\sigma_y} & \lambda'_{x\varphi} \\ \lambda'_{y\sigma_x} & \lambda'_{y\sigma_y} & \lambda'_{y\varphi} \\ \lambda'_{M\sigma_x} & \lambda'_{M\sigma_y} & \lambda'_{M\varphi} \end{bmatrix} \triangleq \begin{bmatrix} \mathbf{C}_\sigma & \mathbf{C}_\varphi \\ \mathbf{C}_{M\sigma} & C_{M\varphi} \end{bmatrix} \begin{bmatrix} \mathbf{S}'_F & \mathbf{S}'_M \\ \mathbf{S}'_{\varphi F} & S'_{\varphi M} \end{bmatrix}. \tag{6.50}$$

To interpret the role of the enhanced relaxation lengths, the following example may be considered.

Example 6.2.1 Neglecting the contribution of the spin slip, for a tyre with isotropic tread, diagonal stiffness matrices, and rectangular contact patch Eq. (6.47) simplifies to

$$\dot{\boldsymbol{F}}_t(t) = V_{\mathrm{r}}(t)\mathbf{C}'_\sigma\Big[\boldsymbol{\sigma}(t) - \hat{\boldsymbol{\sigma}}\big(\boldsymbol{F}_t(t)\big)\Big], \tag{6.51}$$

where the theoretical slip functions $\hat{\boldsymbol{\sigma}}(\cdot)$ read [13]

$$\hat{\boldsymbol{\sigma}}(\boldsymbol{F}_t) = \sigma^{\mathrm{cr}}\left(1 - \sqrt[3]{1 - \frac{F_t}{\mu F_z}}\right)\frac{\boldsymbol{F}_t}{F_t}, \tag{6.52}$$

and the enhanced stiffness matrix $\mathbf{C}'_\sigma$ is given by

$$\mathbf{C}'_\sigma = \begin{bmatrix} C'_{\sigma_x} & 0 \\ 0 & C'_{\sigma_y} \end{bmatrix} \triangleq \begin{bmatrix} \dfrac{C'_x C_\sigma}{aC'_x + C_\sigma} & 0 \\ 0 & \dfrac{C'_y C_\sigma}{aC'_y + C_\sigma} \end{bmatrix}, \tag{6.53}$$

where, as usual, it has been renamed $C'_{x\sigma_x} = C'_{\sigma_x}$ and $C'_{y\sigma_y} = C'_{\sigma_y}$ without ambiguity. Equation (6.51) may be reinterpreted in terms of enhanced relaxation lengths as

$$\mathbf{\Lambda}'_{\sigma}\dot{\boldsymbol{F}}_t(t) = V_{\mathrm{r}}(t)C_{\sigma}\Big[\boldsymbol{\sigma}(t) - \hat{\boldsymbol{\sigma}}\big(\boldsymbol{F}_t(t)\big)\Big], \tag{6.54}$$

with

$$\mathbf{\Lambda}'_{\sigma} = \begin{bmatrix} \lambda'_{\sigma_x} & 0 \\ 0 & \lambda'_{\sigma_y} \end{bmatrix} \triangleq C_{\sigma}\mathbf{C}'^{-1}_{\sigma} = \begin{bmatrix} \dfrac{C_{\sigma}}{C'_{\sigma_x}} & 0 \\ 0 & \dfrac{C_{\sigma}}{C'_{\sigma_y}} \end{bmatrix} = \begin{bmatrix} \dfrac{aC'_x + C_{\sigma}}{C'_x} & 0 \\ 0 & \dfrac{aC'_y + C_{\sigma}}{C'_y} \end{bmatrix}. \tag{6.55}$$

Again, for ease of notation, the diagonal relaxation lengths have been renamed $\lambda'_{x\sigma_x} = \lambda'_{\sigma_x}$ and $\lambda'_{y\sigma_y} = \lambda'_{\sigma_y}$.

The sliding functions $\hat{\boldsymbol{\sigma}}(\cdot)$ in Eq. (6.52) correspond to the steady-state slip-tyre relationships of Subsect. 1.4.3 in absence of spin. These, together with the total translational slip, are shown in Fig. 6.1 as a function of the longitudinal and lateral tyre forces. They may be interpreted as surfaces embedded in a three-dimensional space.

From Eq. (6.55), it is interesting to observe that, when the contact length is small, that is $a \to 0$, the matrix of enhanced relaxation lengths reduces to the conventional one in Eq. (6.11). Basically, the terms aC'_x and aC'_y in the numerators of λ_{σ_x} and λ_{σ_y} in Eq. (6.55) account for the additional term relating to the transient of the bristles. When the contact patch is small, the nonstationary deformation of the bristles becomes negligible compared to that of the tyre carcass, and the two-regime model behaves as the single contact point. Another important aspect concerns the anisotropic properties of the tyre carcass. In this context, a pivotal role is played by the quantity χ_{λ}, defined as the ratio between the smallest and largest relaxation length. Usually, the tyre carcass is much stiffer in the longitudinal direction, that is $C'_x > C'_y$, and therefore $\lambda'_{\sigma_y} > \lambda'_{\sigma_x}$. It may be proved that any value of $\sigma < \sigma_{\chi} \triangleq \lambda_{\chi}\sigma^{\mathrm{cr}}$ ensures that the total planar force never equals nor exceeds the friction limit, that is $F_t(\sigma) < \mu F_z$[1]. In transient conditions, indeed, even values of σ which are lower than the critical value may cause the tyre to fully slide. The values attained in steady-state conditions for different combinations of $\boldsymbol{\sigma}$ such that $\sigma = \sigma_{\chi}$ are shown in the three plots on the left-hand side of Fig. 6.2. The right-hand side plots depict the corresponding values of the total force $F_t(\sigma_{\chi})$ for different values of χ_{λ}. These amount to the 70.4, 92.1 and 98.44% of the peak force.

It should be noticed that the model in Example 6.2.1 is unable to capture the dynamics of the self-aligning moment. This is not, however, an intrinsic limitation

[1] Basically, the interior of the friction circle is an invariant set for $\boldsymbol{F}_t(t)$ [15]. This result may be easily generalised to any model in which the sigma surfaces are are given as $\hat{\boldsymbol{\sigma}}(\boldsymbol{F}_t) = f(F_t)\boldsymbol{F}_t$, where $f(\cdot)$ is a scalar positive function, and the matrix of the enhanced stiffness $\mathbf{C}'_{\sigma}$ is symmetric and positive definite.

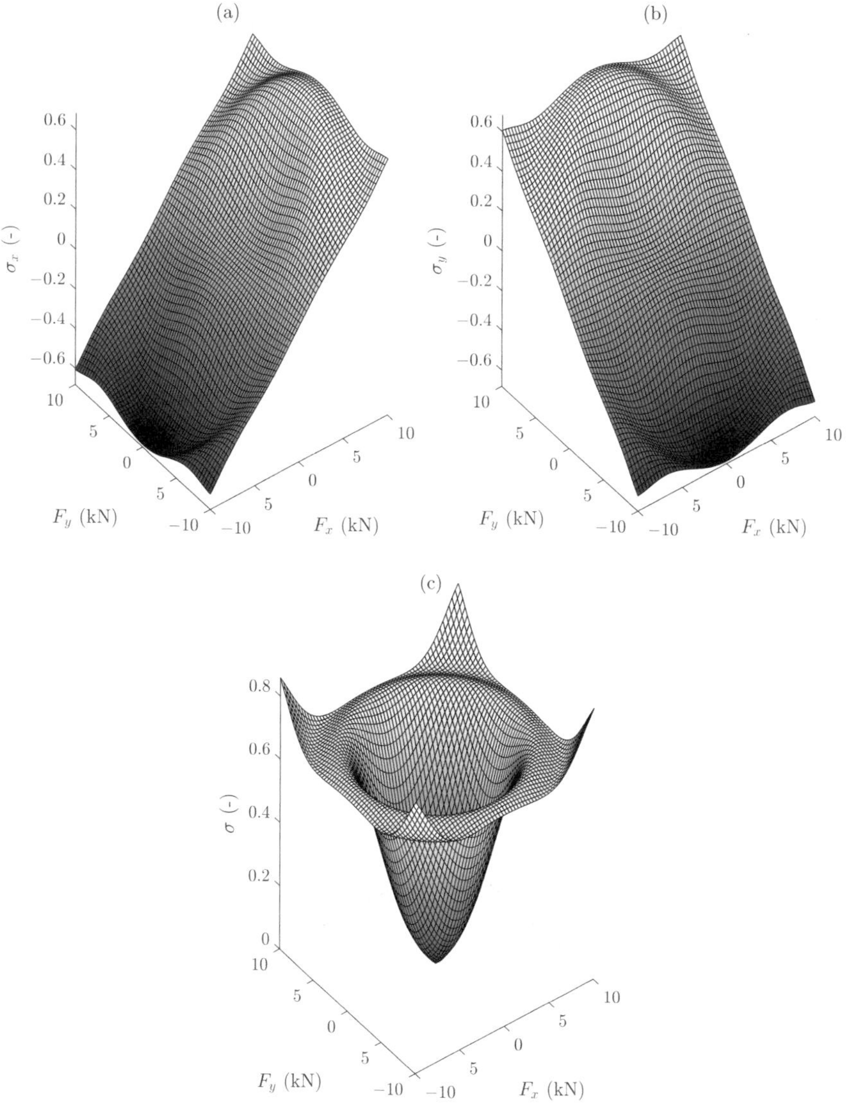

Fig. 6.1 Theoretical slip variables σ_x, σ_y and total slip σ (Subfigures 6.1 (a), (b) and (c), respectively) as a function of the longitudinal and lateral tyre force given as in Eq. (6.52). Tyre parameters: $C_\sigma = 30000$ N, $\mu = 1$, $\sigma^{\text{cr}} = 0.3$, $F_z = 3000$ N

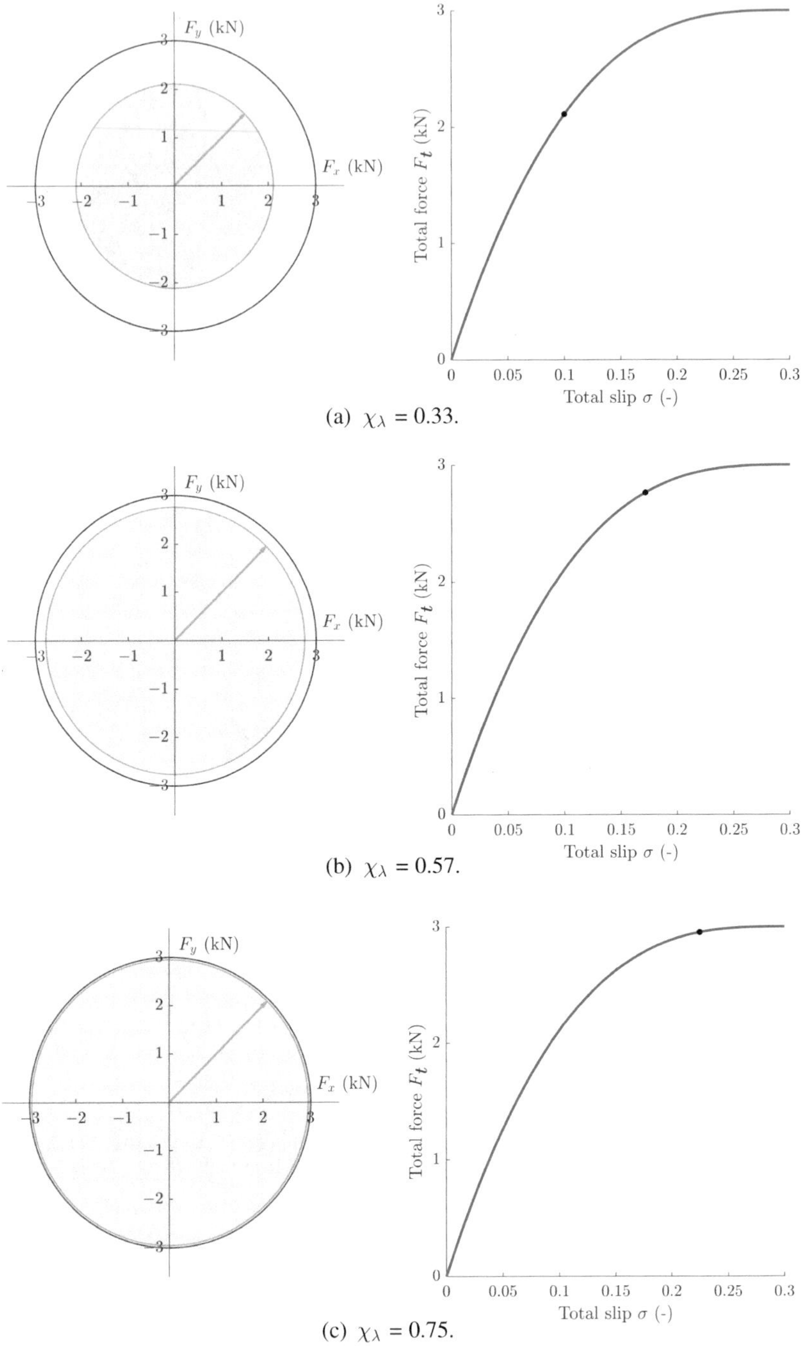

(a) $\chi_\lambda = 0.33$.

(b) $\chi_\lambda = 0.57$.

(c) $\chi_\lambda = 0.75$.

Fig. 6.2 Friction circle and maximum attainable steady-state tangential force for a tyre with anisotropic carcass with different relaxation ratios χ_λ

of the two-regime formulation. In the above example, it is due to the fact that the theoretical slip functions (6.52) have been derived directly from the assumption $\varphi = 0$ from the steady-state relationships (3.31). In this case, there are only two slip variables, and the equation for the self-aligning moment becomes redundant. This aspect may be further clarified considering the next example.

Example 6.2.2 Again for a tyre with isotropic tread and rectangular contact patch, when considering the self-aligning moment, Eq. (6.47) reduces to

$$\begin{bmatrix} \dot{F}_x(t) \\ \dot{F}_y(t) \\ \dot{M}_z(t) \end{bmatrix} = V_{\mathrm{r}}(t) \begin{bmatrix} C'_{\sigma_x} & 0 & 0 \\ 0 & C'_{\sigma_y} & 0 \\ 0 & 0 & C'_{M\varphi} \end{bmatrix} \begin{bmatrix} \sigma_x(t) - \hat{\sigma}_x\big(F_x(t), M_z(t)\big) \\ \sigma_y(t) - \hat{\sigma}_y\big(F_y(t), M_z(t)\big) \\ \varphi(t) - \hat{\varphi}\big(F_y(t), M_z(t)\big) \end{bmatrix}, \tag{6.56}$$

where C'_{σ_x} and C'_{σ_y} read as in Eq. (6.53) and the enhanced aligning spin stiffness is given by

$$C'_{M\varphi} = C_{\varphi} + \frac{C_{M\varphi}}{a}. \tag{6.57}$$

The theoretical slip and spin functions may be obtained by inverting the linear relationships derived under the assumption of vanishing sliding:

$$\begin{bmatrix} \hat{\sigma}_x(F_x, M_z) \\ \hat{\sigma}_y(F_y, M_z) \\ \hat{\varphi}(F_y, M_z) \end{bmatrix} = \begin{bmatrix} \dfrac{1}{C_{\sigma}} & 0 & 0 \\ 0 & \dfrac{C_{M\varphi}}{C_{M\sigma_y} C_{\varphi} + C_{M\varphi} C_{\sigma}} & -\dfrac{C_{\varphi}}{C_{M\sigma_y} C_{\varphi} + C_{M\varphi} C_{\sigma}} \\ 0 & \dfrac{C_{M\sigma_y}}{C_{M\sigma_y} C_{\varphi} + C_{M\varphi} C_{\sigma}} & \dfrac{C_{\sigma}}{C_{M\sigma_y} C_{\varphi} + C_{M\varphi} C_{\sigma}} \end{bmatrix} \begin{bmatrix} F_x \\ F_y \\ M_z \end{bmatrix}. \tag{6.58}$$

It is trivial to verify that the system above is globally asymptotically stable [14, 15]. From Eqs. (6.56) and (6.58), it may be also deduced that the dynamics of the self-aligning moment is affected by both the lateral slip and spin. If the slip and spin functions had been derived by neglecting the spin *a priori*, the last relationship in (6.58) would have become redundant and the two-regime model would not have been able to capture the transient dynamics of the self-aligning moment.

6.2.2.2 Formulation in Terms of Practical Slips

Combining Eqs. (6.33a), (6.33b) and (6.43) according to Eqs. (6.36a), (6.36b) and inverting yields

$$\begin{bmatrix} \dot{\boldsymbol{F}}_t(t) \\ \dot{M}_z(t) \end{bmatrix} = \begin{bmatrix} \mathbf{C}'_{\kappa} & \mathbf{C}'_{\varphi} \\ \mathbf{C}'_{M\kappa} & C'_{M\varphi} \end{bmatrix} \begin{bmatrix} V_{Cx}(t)\Big[\boldsymbol{\kappa}(t) - \hat{\boldsymbol{\kappa}}\big(\boldsymbol{F}_t(t), M_z(t)\big)\Big] \\ V_{\mathrm{r}}(t)\Big[\varphi(t) - \hat{\varphi}\big(\boldsymbol{F}_t(t), M_z(t)\big)\Big] \end{bmatrix}, \tag{6.59}$$

again with

$$\begin{bmatrix} \mathbf{C}'_{\kappa} & \mathbf{C}'_{\varphi} \\ \mathbf{C}'_{M\kappa} & C'_{M\varphi} \end{bmatrix} \triangleq \begin{bmatrix} \mathbf{S}'_{F} & \mathbf{S}'_{M} \\ \mathbf{S}'_{\varphi F} & S'_{\varphi M} \end{bmatrix}^{-1} \equiv \begin{bmatrix} \mathbf{C}'_{\sigma} & \mathbf{C}'_{\varphi} \\ \mathbf{C}'_{M\sigma} & C'_{M\varphi} \end{bmatrix}. \tag{6.60}$$

In analogy to Eq. (6.47), the above relationships (6.59) may be restated in terms of enhanced practical relaxation lengths as follows:

$$\begin{bmatrix} \mathbf{\Lambda}'_{\kappa} & \mathbf{\Lambda}'_{\varphi} \\ \mathbf{\Lambda}'_{M\kappa} & \lambda'_{M\varphi} \end{bmatrix} \begin{bmatrix} \dot{\boldsymbol{F}}_{t}(t) \\ \dot{M}_{z}(t) \end{bmatrix} = \begin{bmatrix} \mathbf{C}_{\kappa} & \mathbf{C}_{\varphi} \\ \mathbf{C}_{M\kappa} & C_{M\varphi} \end{bmatrix} \begin{bmatrix} V_{Cx}(t)\Big[\boldsymbol{\kappa}(t) - \hat{\boldsymbol{\kappa}}\big(\boldsymbol{F}_{t}(t), M_{z}(t)\big)\Big] \\ V_{\mathrm{r}}(t)\Big[\varphi(t) - \hat{\varphi}\big(\boldsymbol{F}_{t}(t), M_{z}(t)\big)\Big] \end{bmatrix} \tag{6.61}$$

with

$$\begin{bmatrix} \mathbf{\Lambda}'_{\kappa} & \mathbf{\Lambda}'_{\varphi} \\ \mathbf{\Lambda}'_{M\kappa} & \lambda'_{M\varphi} \end{bmatrix} = \begin{bmatrix} \lambda'_{x\kappa_x} & \lambda'_{x\kappa_y} & \lambda'_{x\varphi} \\ \lambda'_{y\kappa_x} & \lambda'_{y\kappa_y} & \lambda'_{y\varphi} \\ \lambda'_{M\kappa_x} & \lambda'_{M\kappa_y} & \lambda'_{M\varphi} \end{bmatrix} \triangleq \begin{bmatrix} \mathbf{C}_{\kappa} & \mathbf{C}_{\varphi} \\ \mathbf{C}_{M\kappa} & C_{M\varphi} \end{bmatrix} \begin{bmatrix} \mathbf{S}'_{F} & \mathbf{S}'_{M} \\ \mathbf{S}'_{\varphi F} & S'_{\varphi M} \end{bmatrix} \equiv \begin{bmatrix} \mathbf{\Lambda}'_{\sigma} & \mathbf{\Lambda}'_{\varphi} \\ \mathbf{\Lambda}'_{M\sigma} & \lambda'_{M\varphi} \end{bmatrix}. \tag{6.62}$$

6.2.2.3 Formulation in Terms of Geometrical Slips

Combining Eqs. (6.34a), (6.34b), (6.34c) and (6.46) according to Eqs. (6.37a), (6.37b), (6.37c) and inverting yields

$$\begin{bmatrix} \dot{\boldsymbol{F}}_{t}(t) \\ \dot{M}_{z}(t) \end{bmatrix} = \begin{bmatrix} C'_{x\kappa_x} & C'_{x\alpha} & C'_{x\gamma} \\ C'_{y\kappa_x} & C'_{y\alpha} & C'_{y\gamma} \\ C'_{M\kappa_x} & C'_{M\alpha} & C'_{M\gamma} \end{bmatrix} \begin{bmatrix} V_{Cx}(t)\Big[\kappa_x(t) - \hat{\kappa}_x\big(\boldsymbol{F}_{t}(t), M_{z}(t)\big)\Big] \\ V_{Cx}(t)\Big[\tan\alpha(t) - \tan\hat{\alpha}\big(\boldsymbol{F}_{t}(t), M_{z}(t)\big)\Big] \\ V_{\mathrm{r}}(t)\Big[\sin\gamma(t) - \sin\hat{\gamma}\big(\boldsymbol{F}_{t}(t), M_{z}(t)\big)\Big] \end{bmatrix}, \tag{6.63}$$

with

$$\begin{bmatrix} C'_{x\kappa_x} & C'_{x\alpha} & C'_{x\gamma} \\ C'_{y\kappa_x} & C'_{y\alpha} & C'_{y\gamma} \\ C'_{M\kappa_x} & C'_{M\alpha} & C'_{M\gamma} \end{bmatrix} \triangleq \begin{bmatrix} S'_{xF_x} & S'_{xF_y} & S'_{xM} \\ S'_{yF_x} & S'_{yF_y} & S'_{yM} \\ S'_{\gamma F_x} & S'_{\gamma F_y} & S'_{\gamma M} \end{bmatrix}^{-1}, \tag{6.64}$$

with the camber stiffnesses satisfying

$$\begin{bmatrix} C'_{x\gamma} \\ C'_{y\gamma} \\ C'_{M\gamma} \end{bmatrix} = \frac{1}{R_{\mathrm{r}}}\big(1 - \varepsilon_{\gamma}\big) \begin{bmatrix} C'_{x\varphi} \\ C'_{y\varphi} \\ C'_{M\varphi} \end{bmatrix}. \tag{6.65}$$

In terms of enhanced geometrical relaxation lengths, Eq. (6.63) becomes

$$\begin{bmatrix} \lambda'_{x\kappa_x} & \lambda'_{x\alpha} & \lambda'_{x\gamma} \\ \lambda'_{y\kappa_x} & \lambda'_{y\alpha} & \lambda'_{y\gamma} \\ \lambda'_{M\kappa_x} & \lambda'_{M\alpha} & \lambda'_{M\gamma} \end{bmatrix} \begin{bmatrix} \dot{\boldsymbol{F}}_{\boldsymbol{t}}(t) \\ \dot{M}_z(t) \end{bmatrix} = -\begin{bmatrix} C_{x\kappa_x} & C_{x\alpha} & C_{x\gamma} \\ C_{y\kappa_x} & C_{y\alpha} & C_{y\gamma} \\ C_{M\kappa_x} & C_{M\alpha} & C_{M\gamma} \end{bmatrix} \begin{bmatrix} V_{Cx}(t)\Big[\kappa_x(t) - \hat{\kappa}_x\big(\boldsymbol{F}_{\boldsymbol{t}}(t), M_z(t)\big)\Big] \\ V_{Cx}(t)\Big[\tan\alpha(t) - \tan\hat{\alpha}\big(\boldsymbol{F}_{\boldsymbol{t}}(t), M_z(t)\big)\Big] \\ V_{\mathrm{r}}(t)\Big[\sin\gamma(t) - \sin\hat{\gamma}\big(\boldsymbol{F}_{\boldsymbol{t}}(t), M_z(t)\big)\Big] \end{bmatrix}, \tag{6.66}$$

with

$$\begin{aligned} \begin{bmatrix} \lambda'_{x\kappa_x} & \lambda'_{x\alpha} & \lambda'_{x\gamma} \\ \lambda'_{y\kappa_x} & \lambda'_{y\alpha} & \lambda'_{y\gamma} \\ \lambda'_{M\kappa_x} & \lambda'_{M\alpha} & \lambda'_{M\gamma} \end{bmatrix} &\triangleq \begin{bmatrix} C_{x\kappa_x} & C_{x\alpha} & C_{x\gamma} \\ C_{y\kappa_x} & C_{y\alpha} & C_{y\gamma} \\ C_{M\kappa_x} & C_{M\alpha} & C_{M\gamma} \end{bmatrix} \begin{bmatrix} S'_{xF_x} & S'_{xF_y} & S'_{xM} \\ S'_{yF_x} & S'_{yF_y} & S'_{yM} \\ S'_{\gamma F_x} & S'_{\gamma F_y} & S'_{\gamma M} \end{bmatrix} \\ &\equiv \begin{bmatrix} \boldsymbol{\Lambda}'_{\kappa} & \boldsymbol{\Lambda}'_{\varphi} \\ \boldsymbol{\Lambda}'_{M\kappa} & \lambda'_{M\varphi} \end{bmatrix} \equiv \begin{bmatrix} \boldsymbol{\Lambda}'_{\sigma} & \boldsymbol{\Lambda}'_{\varphi} \\ \boldsymbol{\Lambda}'_{M\sigma} & \lambda'_{M\varphi} \end{bmatrix}. \end{aligned} \tag{6.67}$$

6.3 Model Comparison

To assess their performance, the approximated transient tyre models developed respectively in Sects. 6.1 and 6.2 are compared against each other for different operational conditions. The travelled distance s is also used as an independent variable in place of the time t. By doing so, the problem under consideration becomes independent of the longitudinal and rolling speeds of the tyre.

A first comparison is performed against the nonlinear full contact patch formulation introduced in Sect. 4.3. Pure longitudinal and lateral slip conditions are assumed in turn. The parameters used for the simulation are listed in Table 6.1. Three models

Table 6.1 Tyre parameters

Parameter	Description	Unit	Value
C_σ	Slip stiffness	N	$3 \cdot 10^4$
C'_x	Longitudinal stiffness of the carcass	$\mathrm{N\,m^{-1}}$	$6 \cdot 10^5$
C'_y	Lateral stiffness of the carcass	$\mathrm{N\,m^{-1}}$	$2.4 \cdot 10^5$
F_z	Vertical force	N	3000
λ_{σ_x}	Longitudinal relaxation length	m	0.05
λ_{σ_y}	Lateral relaxation length	m	0.125
a	Contact patch length	m	0.075
μ	Friction coefficient	-	1

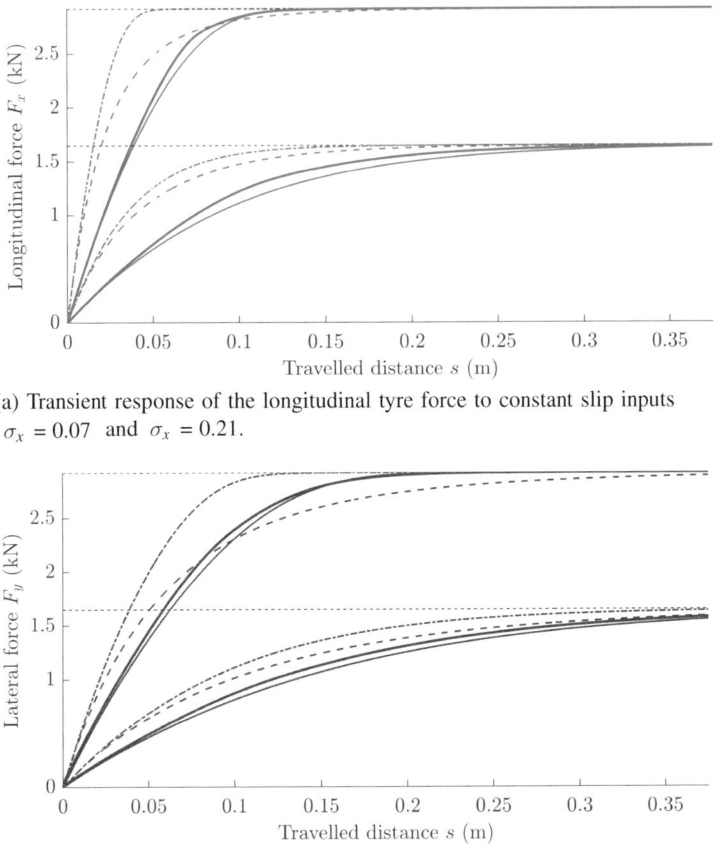

(a) Transient response of the longitudinal tyre force to constant slip inputs $\sigma_x = 0.07$ and $\sigma_x = 0.21$.

(b) Transient response of the lateral tyre force to constant slip inputs $\sigma_y = 0.07$ and $\sigma_y = 0.21$

Fig. 6.3 Comparison between the two-regime tyre model (solid line), the semi-nonlinear (dashed line), the full-nonlinear single contact point model (dash-dotted line) and the nonlinear full contact patch (thick line)

are compared against the exact formulation: the full nonlinear single contact point model, the semi-nonlinear one and the two-regime model for combined slips. These correspond to the models presented in Examples 6.1.1, 6.1.2 and 6.2.1, respectively. Simulation results are shown in Fig. 6.3 for constant slip inputs of $\sigma_x = 0.07$ and 0.21 for the longitudinal case. The same values are used for the lateral interaction. The initial conditions for the tyre forces are assumed to be zero in both cases, that is $\boldsymbol{F}_{t0} = \boldsymbol{F}_t(0) = \boldsymbol{0}$, but other initial conditions are also possible. Generally speaking, it may be observed that the two-regime model succeeds better in replicating the exact trend, whereas the single contact point models exhibit larger discrepancies, especially at higher values of the slip inputs. This phenomenon is particularly evident in the longitudinal case. Indeed, the tyre carcass is stiffer in the longitudinal

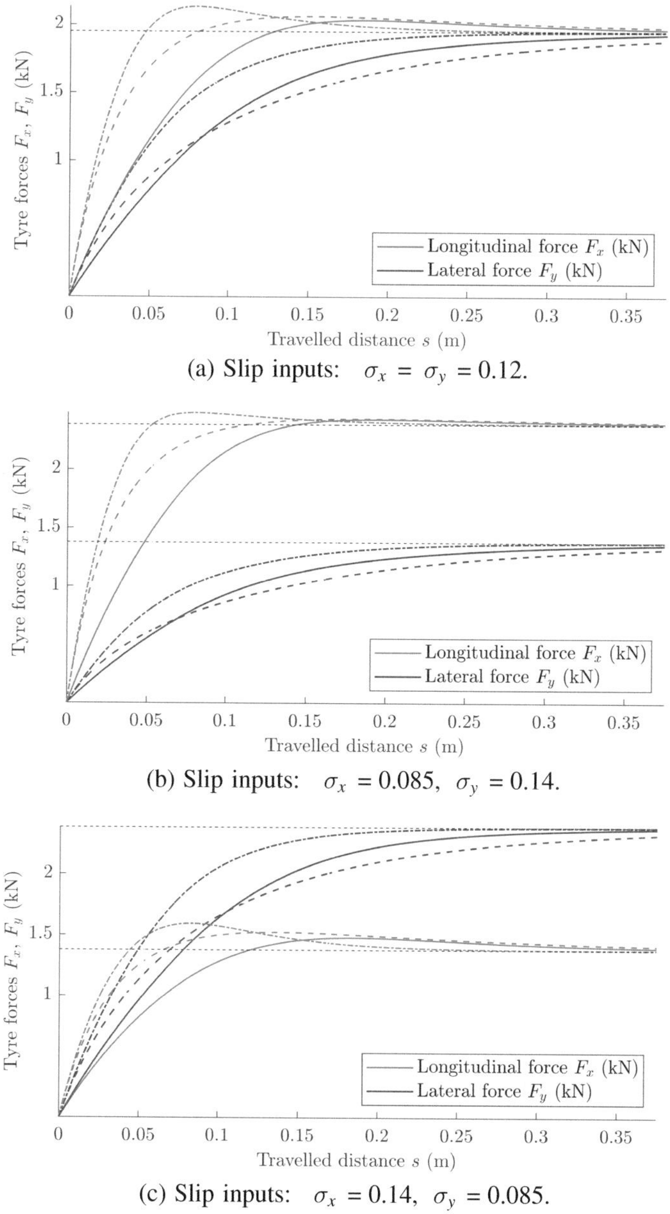

Fig. 6.4 Comparison between the two-regime tyre model (solid line), the semi-nonlinear (dashed line) and the full-nonlinear single contact point model (dash-dotted line) for different combined slip inputs σ_x and σ_y

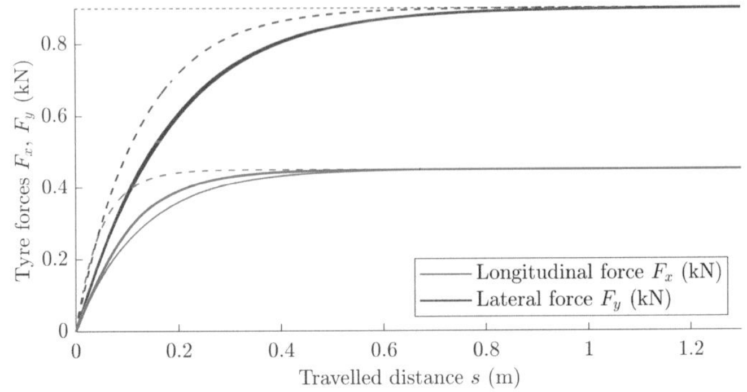

(a) Transient response of the tyre forces to constant slip inputs $\sigma_x = 0.015$ and $\sigma_x = 0.03$.

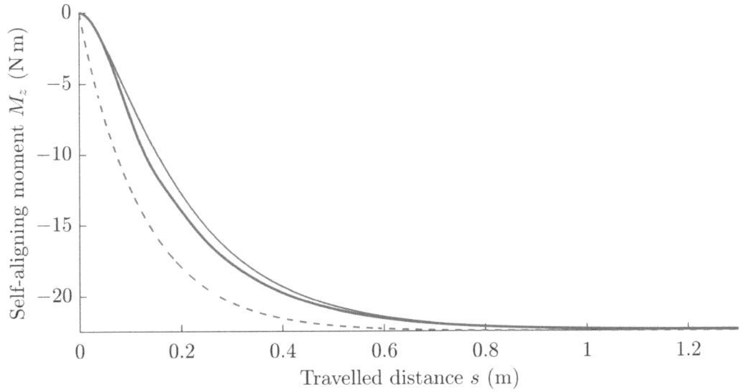

(b) Transient response of the self-aligning moment to constant slip inputs $\sigma_y = 0.015$ and $\sigma_y = 0.03$

Fig. 6.5 Comparison between the linear two-regime tyre model (solid line), the linear single contact point model (dashed line) and the linear full contact patch (thick line)

direction, which implies shorter relaxation lengths. As a consequence, the bristle dynamics and the length of the contact patch play a significant role in determining the transient response of the tyre.

A second comparison between the three approximated models is shown for different combinations of constant slip inputs σ_x and σ_y in Fig. 6.4. In combined slip conditions, the values of the total slip σ need to be considered that are lower than the transient critical value σ_χ. In Fig. 6.4, the total slip amounts to $\sigma = 0.17 < \sigma_\chi$. It may be noticed that the agreement between the two-regime tyre model and the semi-nonlinear single contact point is particularly good in the beginning when the forces are in the linear region of the diagram $\boldsymbol{F}_t - \boldsymbol{\sigma}$ and the assumption of constant relaxation length holds. As the travelled distance increases and the tyre characteristics develop fully, the two-regime transient model exhibits a faster convergence to the steady-state solution, and tends to the full-nonlinear one. Clearly, all the models converge again to the same asymptotic value.

The last set of simulations is aimed at comparing the transient response of the tyre characteristics according to the linear versions of the single contact point and two-regime models, that is the formulations in Examples 6.1.3 and 6.2.2, respectively. The comparison is made again against the nonlinear full contact patch model in Fig. 6.5 starting from an initial undeformed configuration. In this case, however, small translational slips are imposed so that that assumption of vanishing sliding conditions holds. The conclusions are the same as previously, with the two-regime model which slightly outperforms the single contact point.

References

1. Pacejka HB (2012) Tire and vehicle dynamics, 3rd edn. Elsevier/BH, Amsterdam
2. Higuchi A (1997) Transient response of tyres at large wheel slip and camber [doctoral thesis]. Delft
3. Higuchi A, Pacejka HB (1997) The relaxation length concept at large wheel slip and camber. Vehicle Syst Dyn 25(sup001):50–64. Available from: https://doi.org/10.1080/00423119708969644
4. Zegelaar PWA (1998) The dynamic response of tyres to brake torque variations and road unevenesses [doctoral thesis]. Delft. Available from: http://resolver.tudelft.nl/uuid:c623e3fc-b88a-4bec-804a-10bcb7e94124
5. Maurice JP, Berzeri M, Pacejka HB (1999) Pragmatic Tyre model for short wavelength side slip variations. Vehicle Syst Dyn 31(2):65–94. Available from: https://doi.org/10.1076/vesd.31.2.65.2096
6. Guiggiani M (2018) The science of vehicle dynamics, 2nd edn. Springer International, Cham(Switzerland)
7. Svendenius J (2007) Tire modelling and friction estimation [dissertation]. Lund
8. Rill G (2019) Sophisticated but quite simple contact calculation for handling tire models. Multibody Syst Dyn 45:131–153. Available from: https://doi.org/10.1007/s11044-018-9629-4
9. Rill G (2020) Road vehicle dynamics: fundamentals and modeling with MATLAB®. 2nd Ed. CRC Press
10. Shaju A, Pandey AK (2020) Modelling transient response using PAC 2002-based tyre model. Vehicle Syst Dyn. Available from: https://doi.org/10.1080/00423114.2020.1802048
11. Pacejka HB, Besselink IJM (1997) Magic formula tyre model with transient properties. Vehicle Syst Dyn 27(sup001):234–249
12. Romano L, Bruzelius F Jacobson B (2020) Unsteady-state brush theory. Vehicle Syst Dyn, 1–29. Available from: https://doi.org/10.1080/00423114.2020.1774625
13. Bruzelius F, Hjort M, Svendenius J (2014) Validation of a basic combined-slip tyre model for use in friction estimation applications. Proc Instit Mech Eng Part D: J Auto Eng 228(13):1622–1629. Available from: https://doi.org/10.1177/0954407013511797
14. Rugh WJ (2007) Linear system theory, 2nd ed. Johns Hopkins University
15. Khalil HK (2002) Nonlinear systems, 3rd edn. Prentice Hall, Upper Saddle River
16. Romano L, Timpone F, Bruzelius F, Jacobson B. A theoretical investigation on transient tyre slip losses using the brush theory. To appear on to Tire Science and Technology